Education Division

AA008950

Held at the 2023 AIChE Annual Meeting

Orlando, Florida, USA
5-10 November 2023

ISBN: 978-1-7138-9308-0

Printed from e-media with permission by:

Curran Associates, Inc.
57 Morehouse Lane
Red Hook, NY 12571

Some format issues inherent in the e-media version may also appear in this print version.

Printed with permission by Curran Associates, Inc. (2024)

For permission requests, please contact AIChE
at the address below.

AIChE
120 Wall Street, FL 23
New York, NY 10005-4020

Phone: (800) 242-4363
Fax: (203) 775-5177

www.aiche.org

Additional copies of this publication are available from:

Curran Associates, Inc.
57 Morehouse Lane
Red Hook, NY 12571 USA
Phone: 845-758-0400
Fax: 845-758-2633
Email: curran@proceedings.com
Web: www.proceedings.com

TABLE OF CONTENTS

INCLUSIVE CLASSROOM STRATEGIES

EDUCATION DIVISION AWARDS PLENARY (INVITED TALKS)

IN HONOR OF PROF. LIANG-SHIH FAN, THE 2022 RECIPIENT OF THE WARREN K. LEWIS AWARD I (INVITED TALKS)

NEW FACULTY FORUM (INVITED TALKS)

INTRODUCTION TO NSF'S CBET DIVISION AND PROPOSAL WRITING STRATEGIES

NEW FUNDING OPPORTUNITIES AT NSF AND PROGRAM OFFICER PANEL DISCUSSION

POSTER SESSION: CHEMICAL ENGINEERING EDUCATION

GRADUATE EDUCATION AND SUPPORT

PREPARING STUDENTS FOR INDUSTRY

COMPUTING AND SIMULATION IN THE CHE CURRICULUM

HANDS-ON ACTIVITIES FOR LEARNING CHEME

SOCIAL JUSTICE AND SOCIETAL IMPACT IN CHEMICAL ENGINEERING

SUSTAINABILITY AND GLOBAL IMPACT IN THE CHE CLASSROOM

ENGAGED TEACHING AND LEARNING

FACULTY PROFESSIONAL DEVELOPMENT - TALKS AND WORKSHOP

STUDENT SUPPORT AND MENTORING

ENTREPRENEURIAL MINDSET AND CHEMICAL ENGINEERING (INVITED TALKS)

UNIT OPERATIONS LABORATORY

TEACHING TRANSPORT PHENOMENA

IN HONOR OF PROF. LIANG-SHIH FAN, THE 2022 RECIPIENT OF THE WARREN K. LEWIS AWARD II (INVITED TALKS)

517a Inclusive Teaching Strategies for Faculty to Improve Student Success

Sandy Pettit, Molly A. Skinner

Inclusive teaching can be described in many ways. The narrative varies due to the framing by contextual factors associated with the program or institution. At its core, the two fundamental elements that emerge are equity and belonging. Inclusive teaching includes the creation of a teaching plan that recognizes and accounts for the diversity of the learners and strives to achieve equivalent learning outcomes for all. Further, it creates a welcoming learning environment that fosters a sense a belonging. There is a natural connection between the use of inclusive pedagogy and improved student success. While there are no one-size-fits-all solutions (due to the contextual factors of the institution and cohort), there are evidence-based strategies that work across many situations. Individual faculty members can employee classroom management techniques to build a sense of community and engineering identity. Additionally, they can use multimodal content and active learning scenarios in their teaching plans. This session will discuss the core elements of inclusive teaching and specific strategies that faculty can use in their course planning and execution.

517b Promotion of an Inclusive Classroom in the Chemical Engineering Program

Nagma Zerin

When students feel comfortable to be who they are and feel that they are part of a supportive environment, it has a positive impact on their learning, academic accomplishments, and retention in their undergraduate program. Chemical Engineering program in general has the reputation of being difficult and highly competitive. It is inevitable for the students to feel overwhelmed and disconnected while handling all the stress and the intensity of the program. Without the presence of sufficient support, the students might suffer from mental health issues, fall behind in their studies, and feel inclined to drop out of the program. Thus, it is imperative for the professors to spend some time to create a classroom environment that not only provides high-quality education but also considers the overall well-being of the students. Some initiatives have been implemented in an elective Chemical Engineering course in our institution, typically taken by sophomores and juniors. Some examples of the initiatives are incorporating a diversity, equity, and inclusion (DEI) statement in the course syllabus, emphasizing the importance of an inclusive classroom at the beginning of the semester, pausing multiple times during a lecture to provide the students sufficient time to absorb the lecture content, providing recorded lectures to accommodate all students' learning pace, creating opportunities for developing connections with the peers through group discussions and group project, having lower weighted exams, and checking on the students to make sure they are doing okay, especially during the exams. A research study is currently being conducted in this course to assess the impact of these initiatives on the students' sense of belonging in the Chemical Engineering major.

517c Deadlines and Ungrading: The Impact of Flexibility on Student Stress and Productivity

Luke Landherr

Ungrading provides an approach that shifts the emphasis of each course onto learning and what students should be able to do at the end of a course, as opposed to focusing students' mindsets on what they need to do to get a certain grade and defining success by the grade they have received. Grades are de-emphasized for greater focus on how well students have learned, processed, and applied the instruction in the course, through means such as greater levels of feedback and discussion with each other and the instructor. There are many approaches to ungrading, such as contract grading, mastery-based grading, portfolios, and self-assessment, but the broader approach allows an instructor to provide less stressful, more equitable assessment.

Additionally, studies have shown the importance of allowing flexibile deadlines with respect to the submission of student work. Students can benefit from having greater means of balancing their work and their personal lives, and the increased quality of life will enable them to achieve greater levels of success in their work and learn more from their academic studies. Flexible deadlines can also be effectively implemented through and ungrading approach, providing an additional approach to assessment that can further improve the student experience and decrease student stress and anxiety.

In a process control course and a mass and energy balances course, an ungrading approach was implemented allowing students to self-determine the assignments they would complete and how many assignments to complete, with all student deliverables culminating in one portfolio of demonstrated work by the end of the semester. Exams were eliminated in favor of the portfolio approach, and deadlines were reduced and made flexible throughout the course. Students were highly productive, completing a greater number of assignments correctly than the number of assignments that would have even been assigned in a more traditionally assessed semester. However, student feedback indicated that having minimal deadlines led to procrastination and was stress-inducing in its own way. In more recent iterations of both courses, students were given different intermediate deadlines for work to be submitted throughout the semester for preliminary feedback, in an effort to study the impact of varying flexibility on students' productivity. The results of these efforts over three semesters, analysis, and student feedback, will be discussed in this paper.

517g Professional Identity and Community Development with the Story Collider

Eric Jankowski

Persistence of students in STEM programs is linked to students' identification with their profession and with their sense of belonging. In this work we describe personal storytelling interventions developed with The Story Collider, a national storytelling non-profit. These curricular interventions arose from a need to meet ABET Outcome 3 (an ability to communicate effectively with a range of audiences) in a Thermodynamics course, and have grown into a multidisciplinary collaboration measuring the impact in belonging, persistence, and student support networks that result from the writing and sharing of personal stories related to ones' major. We describe quantitative shifts in attitudes related to confidence in subject material and identification with ones' major, and discuss qualitative correlations with the themes that emerge from student stories and in anonymized interviews. We conclude with a practical menu of classroom activities from small Scientist Spotlights and medium-sized story-writing activities that can be applied across undergraduate and graduate engineering curricula.

This material is based upon work supported by the National Science Foundation under Grant IUSE-2142137.

517h My Professor Listens to Rage Against the Machine? Using No-Stakes Personal Trivia Questions to Build Student-Teacher Relationships

Matthew Cooper

Interactive polling software tools such as Top Hat have enabled entire classrooms of students to rapidly provide answers to instructor questions. This practice has many advantages, particularly by enhancing just-in-time teaching through allowing faculty to identify areas in which students have muddy understanding of course concepts. However, these software tools can also be used for more informal endeavors not driven by course content. This work describes using polling software to communicate personal information about the course instructor through a series of questions throughout the semester, with the goals of increasing student engagement and allowing students to view their instructor as a "real person," thus improving student-teacher relationships. In a sophomore numerical methods course taught by the author, a 5-minute break was regularly provided in the middle of each 75-minute T/Th lecture period. On each Thursday throughout the semester when no exam was scheduled, this break included a single no-stakes polling question regarding the course instructor's personal life. For instance, one question asked students to select which band (out of four choices) performed at the only music concert the instructor attended in 2022. Once the answer to the question was revealed, the instructor provided brief commentary on the correct answer, poll results, etc. Since the instructor writes all questions, the personal information shared with students is at their discretion. This talk will share example questions and tips for faculty interested in the practice. Further, in order to understand student perspectives on this practice a question was included in end-of-semester course evaluations which asked students whether they felt these polling activities were worthwhile and if they recommend the instructor continue the practice in future semesters. A brief qualitative investigation via thematic analysis of responses will be presented to illustrate prevailing student perspectives.

46a Navigating Professional Service When It's Personal

Anthony Butterfield

Presentation by the recipient of the 2021 Award for Service to Chemical Engineering Education

"For dedicated leadership and service in chemical engineering education, including contributions toward diversity and equity, particularly fostering an inclusive climate for LGBTQ+&Ally students and professionals."

Building on the contributions of many who came before them, an engineer can see even their small improvements to a process lead to wide-reaching positive impacts. This potential of chemical engineering to be a potent helping profession not only attracts new engineers but also serves as an addictive source of purpose on the job. To build on the work of others and find those small improvements with large impact similarly motivates a great deal of professional service, particularly for those who have experienced problematic communities in the past.

For that motivation, many of us say 'yes' to every service opportunity that comes our way, reflexively. There is a lot to say 'yes' to: advising student groups, reviewing papers and proposals, chairing divisions, joining committees, giving workshops, chairing committees, community outreach, initiating initiatives, and much more. Additionally, there is a lot of community service one can not reasonably say 'no' to, particularly as an educator, from pushing back against bias to helping students in distress connect with the help they need. The emotional and social labors can eclipse the technical labors, and an engineer could work through each day on only service to their community, without ever using the technical skills of their formal education.

I have taken on a fair share of those professional service opportunities. I have done a fair percentage of it acceptably, but stumbled at an equally fair percentage. In this talk, I aim to share the professional service insights I wish I had encountered earlier in my career. Additional focus will be given to professional service around the LGBTQ+ engineering community, and I will discuss the complicating role that personal histories with bias in engineering can have in compelling and contextualizing service roles.

46b Cooperative, Hands-on, Active, Problem-Based Learning – Ideas and Insights for the Classroom

Bernard Van Wie

Presentation by the recipient of the 2022 Award for Innovation in Chemical Engineering Education.

"For the development, deployment and assessment of innovative desktop learning modules enabling an experiential understanding of fundamental engineering phenomena implemented in a traditional classroom setting."

128a Professor L.-S. Fan's Career and Legacy As a Researcher, Educator, and Mentor

Ah-Hyung Alissa Park, Robert Pfeffer

It is not hyperbole to assert that Prof. Fan is one of the intellectual leaders of Chemical Engineering. He is undoubtedly one of the most celebrated and respected Chemical Engineers in the world. He is a member of the NAE, six foreign academies, and the National Academy of Inventors. He was also chosen by the AIChE as one of the "One Hundred Chemical Engineers of the Modern Era." We believe that his impact on sustainable energy has been far beyond any other academic researchers of recent years.

More importantly, Prof. Fan is an amazing teacher and mentor. He truly enjoys teaching and mentoring, and takes a great pride in his profession as a teacher. He spends amazing efforts in preparing, refining and updating lectures even for courses that he has taught for many years. He continues to reinvent his courses and introduce new teaching modules and tools to provide better learning experiences for all his students. Undergraduate and graduate students and faculty have recognized his outstanding teaching ability and pedagogical skills and have honored him with a number of Department, College and University distinguished teaching awards. His over four decades of tremendous educational impacts have been propagated throughout our Chemical Engineering profession. With his devoted effortseducating and mentoring young faculty and students with different challenges, the Diversity, Equity and Inclusion (DE&I) in Chemical Engineering has been greatly improved.

Prof. Fan has influenced a number of areas in Chemical Engineering research and education including sustainable energy and particle technology. Prof. Fan's redox chemical looping reaction can produce syngas, methanol, H_2, and liquid fuels from various carbonaceous feedstock, while achieving CO_2 neutrality or even carbon negativity. For a century, industries believed it was impossible to synthesize metal oxide oxygen carriers with sustainable redox properties for any meaningful commercial applications. Prof. Fan's breakthrough revealed the critical metal oxide redox mechanism underlining the ionic migration and hence its control, leading to robust redox particle formulation and synthesis which has entirely changed that perception making a dream is now reality. His metal oxide oxygen carriers have been tested for over 10,000 redox cycles and operated beyond the 'Holy Grail' of particle formulations.

Now, for the first time, the classical chemical reactions used over the past 50 years, i.e., autothermal and steam methane reforming processes, can be replaced. The impact of his chemical looping technology is tremendous. Professor Fan's remarkable innovations in chemical looping are just one important facet of his prolific scholarly legacy, and it will serve as one of the foundations for our future Chemical Engineering education. Thus, it is a great honor to celebrate Prof. L.-S. Fan's career and legacy as a researcher, educator, and mentor with the 2022 AIChE Warren K. Lewis Award for Chemical Engineering Education.

128b *Excellence in Engineering Education and Research - Perspectives from Former Students and Postdocs of Prof. L.S. Fan*

Chao Zhu, Xiaotao Bi, Raymond Lau, Mandar Kathe, Fanxing Li

This presentation brings together five former PhD. students and postdoctoral researchers of Prof. L.S. Fan, the recipient of the 2022 Warren K. Lewis Award for Chemical Engineering Education. Through his mentorship, Professor Fan has had a significant impact on the careers of many successful engineers across multiple disciplines. In our presentation, we will share our individual perspectives on his unique approach to engineering education and research. We will reflect on the key lessons we learned from him and discuss the ways in which his mentorship has shaped our own academic careers. Our presentation will also highlight how we have incorporated his teachings into our own work, and how we have benefited from his guidance in our professional and personal lives. Ultimately, we hope to inspire the next generation of engineers to strive for excellence and contribute to the advancement of the field.

128c Contact Charging in Particulate Flows

Sankaran Sundaresan

Contact charging is a well-known phenomenon wherein material surfaces become asymmetrically charged following their contact and separation. In particulate flows, contact charging occurs through a combination of particle-particle and particle-wall interactions. For insulating materials, water is thought to play a central role, through the transport of water ions. This talk will first touch upon recent molecular dynamics simulations to characterize the thermodynamic driving forces for water ions to transfer from one surface to another [1]. It will then examine the effects of space charge and micro-discharges in the gas on the extent of particle charging [2-4]. It will then illustrate the interplay between gas-particle flows and charge accumulation on particles in fluidized beds.

[1] M.A. Webb, H. Zhang, and S. Sundaresan, "Can thermodynamics explain contact charging between insulating polymer surfaces?", paper presented at the APS March Meeting 2023, Las Vegas, NV.

[2] X. Liu, J. Kolehmainen, I. Nwogbaga, A. Ozel, and S. Sundaresan, "Effect of particle size on tribocharging," *Powder Technol.*, **375**, 199-209 (2020)

[3] X. Liu and S. Sundaresan, "The effect of gas on tribocharging of particles in a vibrated bed," *Powder Technol.*, **401**, 117272 (2022).

[4] J. Hou, X. Liu, M. H. Rahaman, H. Li, S. Sundaresan, J. R. Grace, and X. Bi, "Effect of gas properties and wall materials on particle charging in fluidized beds," *Can. J. Chem. Eng.*, **101**, 244-255 (2023).

128d LS Fan: Inspiring Educator and Innovator

Marc-Olivier Coppens

I will present a short overview of our work on nature-inspired chemical engineering (NICE) as a method to dynamically structure fluidized beds, harnessing fundamental insights from scalability and pattern formation in natural systems to apply to gas-solid fluidization with its many industrial applications. Parallel to this overview, I will discuss the impact of Professor LS Fan on me as an engineer, researcher, educator and in leadership positions over the past 25 years, as my own career developed from TU Delft to Rensselaer and UCL. Along the way, LS Fan's impact evolved from encouragement to my early ideas as starting researcher in fluidization; lessons from his educational research articles, books, and presentations; work on AIChE committees; to insights on commercialisation (from his impressive work on scaling up chemical looping) and on topics well beyond fluidization, such as approaches to academic leadership. Every conversation was edifying. Mentorship evolved into friendship. LS Fan's modesty and humility through his listening as well as thoughtful advice given to me as someone who was never formally his student, but also others I have been able to witness, are testament to his intellectual and emotional intelligence and skills as selfless educator and mentor.

128e Status of Particle Technology Education in the United States

Jennifer Curtis

The majority of products produced in the chemical process industries either involve particles in their processing or are in particulate form in their final state. Lack of knowledge of how particles behave differently than fluids often results in a loss of production, poor product quality, and failures in process and equipment design. Despite the pervasiveness of particles in processes, the extent of particle technology education in typical chemical engineering programs in the US is lacking. However, efforts have been made to address this gap in particle technology education in terms of courses, textbooks and other educational materials, and even degree programs. This talk will review the current status of particle technology education in the US.

128f My Association with LS Fan: LB-DEM Study of Particle-Fluid Interaction Forces As an Example

Aibing Yu

This talk will first briefly review my association with Professor LS Fan in the past 25 years or so, demonstrating how I knew him from a researcher to collaborator, mentor and finally a good friend. Then, using the LB-DEM study of particle-fluid interaction forces as an example to demonstrate how his research has led to a solution to a significant problem where the drag force has been determined from single- to multi-particles, from mono- to multis-sized particles, from spherical to non-spherical (ellipsoidal) particles, and from Newtonian to Non-Newtonian fluids.

128g Exceptional Contributions As a Green Engineering Educator and Innovator in Taiwan: L.-S. Fan

Pen-Chi Chiang, Shu-Yuan Pan

This presentation highlights the exceptional contributions of Professor L.-S. Fan as a green engineering educator and innovator in Taiwan. Throughout his career, Fan has demonstrated an unwavering commitment to education and has mentored faculty and students at various institutions, including National Taiwan University and National Taiwan University of Science and Technology. His creative ability has resulted in numerous scholarly contributions to literature and inventions that have made a significant impact on the field of green engineering in Taiwan. It is worth noting that Professor Fan's advice on CO2 mineralization technologies in Taiwan has led to several SCI publications, including Nature Sustainability and Environmental Science & Technology. These technologies have also been tested in local industries, such as China Steel Corp., demonstrating their practical application and impact in the field of green engineering. In addition to his work in education, Professor Fan has contributed to the development and deployment of green engineering in industry through consulting and government service in Taiwan. He is also recognized as an academic leader and innovator, not only in Taiwan but throughout the world. Fan's many years of research and education have led to significant contributions in the science and practice of green science, technologies, and engineering for environmental sustainability. He is particularly well-known for his contributions to chemical looping technology, which have garnered recognition globally. Beyond his outstanding achievements in education and research, Professor Fan has also served as a member of the international advisory committee for the Carbon Cycle Research Center (CCRC) at National Taiwan University for over a decade. He has continuously organized the Green Energy Conference in Taiwan every year, facilitating the exchange of information on clean energy and CO2 capture and utilization technologies. Overall, Professor L.-S. Fan's contributions as a green engineering educator and innovator in Taiwan have been exemplary, and his impact on the field has been felt globally. His work has been instrumental in advancing the field of green engineering and promoting environmental sustainability in Taiwan and beyond.

5a Becoming an Assistant Professor

Sophia Orbach

In this session, I will discuss my journey in obtaining a tenure-track position. I will share my experiences as a postdoc, the impact of COVID, the application process, and the benefits of participating in AIChE-sponsored programs such as the Meet the Faculty Candidates Poster Session, Future Faculty Mentoring Program, and Summer School for Engineering Faculty. I will discuss the transition in responsibilities and expectations from postdoc to a new assistant professor as well as the transition from degrees in chemical engineering to research in biomedical engineering.

5b Transition in Progress

Damilola Daramola

In this session, I will share my path to becoming tenure-track faculty and the multiple transitions along the way that have included being an industrial intern, an instructor, an administrator, and subsequently changing institutions prior to obtaining tenure. This conversation will offer an opportunity to the unique path that also included finding/working with appropriate mentors, a supportive environment, and empowering your young researchers to do the same.

5c Learning on the Job: Launching an Academic Career

Karthish Manthiram

I will share strategies for successfully pursuing, evolving, and refining the vision that brought you to academia. This will intersect with topics regarding group management, including how to empower group members to feel ownership over the way in which the group is run and how to experiment not just in your science but in the operation of a group. Many of the changes that happen upon starting a junior faculty career involve learning on the job – doing things that you were not trained to do, and we will discuss strategies for adapting to a new set of circumstances. Mechanisms for evolving in the role will be presented as well. Overall, the strategies outlined here will provide a path to not just advancing your scientific vision but staying true to the values you hold inside and outside academia.

5d Making Your Research Matter: Strategies for Successful Broader Impacts Statements

Sindia M. Rivera-Jimenez, Sarah Wilson

This oral presentation is aimed at new faculty starting their careers and focuses on best practices for writing a strong Broader Impacts Statement for NSF proposals. The presentation emphasizes the importance of considering the impact of scientific research on society beyond the lab and highlights the role of the Broader Impacts Statement as a crucial aspect of any research proposal submitted to NSF. The presentation identifies various strategies for creating a successful Broader Impacts Statement, including identifying desired societal outcomes, selecting relevant evidence-based practices for dissemination, finding collaborators, and incorporating the Broader Impacts Impact Framework. Additionally, the presentation emphasizes the importance of evaluating the impact of proposed educational activities, and accounting for the time and costs associated with broader impacts activities. It also covers common pitfalls to avoid in writing a Broader Impacts Statement, such as a generic or unrealistic educational component, reinventing the wheel with an outreach deliverable, and focusing solely on education. The presentation is intended to provide new faculty with practical guidance for writing a strong Broader Impacts Statement that advances science and society while increasing the fundability of NSF proposals.

16a Division Director Update on the Chemical, Bioengineering, Environmental, and Transport Systems Division of NSF: Programs and Opportunities

Jeanne VanBriesen

The mission of the Chemical, Bioengineering, Environmental and Transport Systems (CBET) Division of NSF's Directorate for Engineering is to support innovative research and education in the fields of chemical engineering, biotechnology, bioengineering, and environmental engineering, and in areas that involve the transformation and/or transport of matter and energy by chemical, thermal, or mechanical means. CBET research and education investments contribute significantly to the knowledge base and to the development of a 21st century workforce for major components of the U.S. economy, including chemicals, pharmaceuticals, medical devices, forest products, metals, petroleum, natural gas, food, textiles, energy utilities, alternative energy sources, microelectronics, and other sectors. Support for environmental research encompasses pollution prevention and remediation as well as life-cycle analysis.

The Division has four clusters of core programs for unsolicited proposals. The Chemical Process Systems (CPS) cluster of programs in CBET supports research seeking innovations in catalytic design, reaction engineering, multiphase chemical separations processes, molecular thermodynamics, process control and design, and sustainable energy conversion. The four programs are Catalysis, Interfacial Engineering, Electrochemical Systems, and Process Systems, Reaction Engineering, and Molecular Thermodynamics. The Transport Phenomena Cluster at CBET supports fundamental research on the transport of mass, momentum, energy, and species. There are four core programs in this cluster: Fluid Dynamics, Particulate and Multiphase Processes, Combustion and Fire systems, and Thermal Transport Processes. Research proposals supported by these core programs span across several disciplines including Chemical, Mechanical, Aerospace, Petroleum, and Environmental Engineering. The Engineering Biology and Health (EBH) cluster of programs in CBET supports a diverse research portfolio, ranging from nano-scale molecular design to macroscale assistive device development. Core programs in the EBH cluster include Cellular and Biochemical Engineering, Biosensing, Biophotonics, Engineering of Biomedical Systems, and Disability and Rehabilitation Engineering. The Environmental Engineering and Sustainability (EES) Cluster is organized around three core programs: 1) Environmental Engineering, 2) Environmental Sustainability, and 3) Nanoscale Interactions.

16b Proposal Writing Strategies for Fundamental Research Proposals for NSF

Gregory L. Rorrer

This workshop, conducted by a former program officer of NSF's Engineering Directorate, will discuss strategies for putting together fundamental research proposals, typical of the National Science Foundation. The workshop will discuss the importance of motivation of the problem, research approach, and describing the intellectual merit and broader impacts of the project. The workshop will be followed with informal networking with NSF program officers at the meeting.

120a New Funding Opportunities at NSF of Interest to the Aiche Research Community

Jeanne VanBriesen

This workshop will highlight Engineering Directorate and cross Directorate funding opportunities at the National Science Foundation that may be of interest to the AIChE Research communities. We will also discuss programs of NSF's newest directorate, Technology, Innovation and Partnerships (TIP). Example programs including Future Manufacturing, Accelerating Innovations in Biomanufacturing with the DOE's Agile BioFoundry, the NSF/CASIS Collaboration on Transport Phenomena Research on the International Space Station.

Process Intensification Approach in Learning, Teaching and Assessment: Course Design and Content Delivery

Jimoh Adewole

Sustainable industrial growth is globally recognised as one of the current challenges of our time. Outcome-based quality education could be one of the solutions to overcoming this challenge. Process intensification (PI) is described as a design approach that can substantially lead to shrinking process equipment size, energy saving, costs reduction, increased safety, and reduced environmental impact. It could also be useful in designing a chemical engineering (ChE) course. The application of best PI practices is based on four principles: maximizing effectiveness of intra & inter molecular events; giving each molecule same processing experience; optimizing the driving forces/resistances at every scale and maximizing the specific surface areas to which these forces /resistances apply; and maximizing the synergistic effect from partial processes. Furthermore, enhanced specific surface area can be realised via structuring the equipment or material used for the process. Moreover, PI can be achieved through the functional, spatial, thermodynamics and temporal domains. Each of these domains has its own peculiarity. For example, in the spatial domain, emphasis is laid on the need to avoid randomness in process equipment. For this reason, it is highly recommended to replace randomness in process equipment with structured ones. Random character at any scale can reduce the efficiency, cause safety problems, reduce process predictability and ability to control process behaviour. One of the major causes of randomness in process equipment or material is the manufacturing technique used for its production. Thus, the intended excellence in chemical engineering education can be achieved by following a systematically structured approach in the course design, content delivery and constructive alignment of the delivered content with the assessment given to students. This work examines how some of the concepts of PI could be applied to develop an effective outcome-based ChE education. It will also provide a case study of the application of PI using a core ChE course, and highlight the challenges involved in adopting this approach.

379a Quantifying Spreadsheet Skills Using an Interactive Textbook

Matthew Liberatore, Grant Valentine

While chemical engineering students and practicing chemical engineers use spreadsheets for completing calculations and many other tasks, learning, practicing, and retaining new spreadsheet skills can be difficult to document. For example, applying a spreadsheet formula for the first time could include demonstration by an expert on YouTube. Spreadsheet programs are easily accessible and have been available for many years, including Microsoft Excel and Google Sheets. While most spreadsheet tasks are the same across platform and application, training on a specific operating system and version is unnecessary. Thus, moving from training to more educational presentation of spreadsheet skills that apply broadly and are transferrable is possible with an interactive textbook. Here, an interactive textbook combines interactive reading content with the opportunity to practice spreadsheet skills through auto-graded problems. The interactive textbook applies learning theories, including cognitive load, scaffolding, and deliberate practice, to both engage and assess learners. First, reading participation is measured by clicks within animations, multiple-choice and true/false questions, and matching exercises. Next, students have unlimited attempts to correctly answer over 140 auto-graded, randomized problems. In addition to the basics of spreadsheeting, advanced topics include solver, error, statistics, and lookup functions. Five cohorts representing over 400 students were studied. Median reading participation was over 96% for all cohorts. In addition, animation view rates were as high as 118%. Animation view rates were highest for more advanced topics, such as double interpolation. High auto-graded problem completion was observed. Median completion over 94% for each cohort was measured. For example, the interpolation section had the lowest mean fraction correct and highest median attempts before correct. By examining fraction correct and attempts before correct at the individual section level, real-time misconceptions and struggle can be noted, which leads to opportunities to provide interventions and facilitate learning. Hypothesis testing and other statistical measures will be presented for both reading participation and auto-graded problems.

379b Matlab Live Scripts for Teaching and Evaluating Fundamental Process Control Concepts

Edward Gatzke

MATLAB introduced Live Scripts in 2016. Live Scripts add a variety of features to traditional MATLAB programming code files, including formatted text blocks and interactive controls. These features offer an opportunity to improve the education of undergraduate students through experiential learning. Using Live Scripts, students can read through detailed equations formatted using LaTeX. Mathematical concepts can be illustrated with inline plots visualizing the equations. User interface controls can be added to the Live Script to allow the user to modify the plot, actively seeing the results of their actions and how a parameter change or value selection impacts the resulting plot. These interactive plots can be used by a new user without any knowledge of the underlying MATLAB code. Additionally, the script can submit online scores so student comprehension can be evaluated efficiently.

This work presents multiple examples of Live Scripts for illustration of selected basic process control concepts. These topics considered include Laplace transform of functions, adding simple functions in the time domain, and step-response FOTD modeling. The Live Scripts can also be used to submit student information for the exercise to a Google Docs spreadsheet. Student response to use of these Live Scripts for education and evaluation will be presented.

379e Girls Talk Math: Redesigning a Curriculum for High Schoolers on Renewable Energy and Recycling

Madison Reed, Francesca Bernardi

Girls Talk Math (GTM) is a free two-week summer camp with the goal of building a welcoming environment for high school girls and non-binary students interested in STEM. Some of the topics students learn include knot theory, quantum mechanics, RSA encryption cryptography, network science, and renewable energy. The renewable energy curriculum was introduced in 2022 at two of the program locations: the University of North Carolina at Chapel Hill and Worcester Polytechnic Institute. Results from post-camp surveys suggested the need for a curriculum update for 2023 to streamline content and include more hands-on activities and less reading. The curriculum has been expanded to include basic chemistry, polymer science, and plastics recycling. Additionally, the problem set has been reworked to be more interactive and relatable to students and their communities. Students also learn about renewable energy sources and build and program their own model windmill using LittleBits circuits and cardboard to power an LED light. In summer 2023, campers will complete post-camp surveys with curriculum-specific questions. By adding quantifiable and relatable examples of chemistry applied in their everyday lives, we hypothesize that students will be more likely to remain engaged with the material and retain the information presented.

379f Integration of Research in Chemical Engineering Education through Service Learning

Julie Albert, Katie Russell, Franziska Trautmann

During the Spring 2021 semester, Prof. Katie Russell partnered with Glass Half Full (GHF), a glass recycling start-up company in New Orleans co-founded by two Tulane students in 2020, for several service-learning projects designed to harness the technical and critical thinking skills of Tulane chemical engineering students. One team experimentally analyzed contamination levels in recycled glass under Prof. Julie Albert's guidance, and their results inspired Albert, Russell, and GHF co-founder Franziska Trautmann (BSE Chemical Engineering, 2020) to assemble a team in response to the National Science Foundation's Convergence Accelerator Program call for proposals related to the networked blue economy (Track E). Now a $5.7 million research effort involving over 20 faculty researchers and a dozen community partnerships, the chemical engineering service-learning projects have grown, engaging freshmen and juniors in projects related to water management (e.g., storm drain filters, French drains and rain gardens) and public education around glass recycling and combatting coastal land loss (e.g., Tik-tok videos). This poster will present the results of these projects as well as anecdotal evidence of their positive impact on recruiting and retaining students in the chemical engineering discipline at a time when enrollment in chemical engineering departments is declining nationwide.

*Funding: NSF 2137730, 2230769

379g Rejuvenation of Old Equipment Using Low-Cost, Ease-of-Use Single-Board Computer to Enrich Process Controls Curriculum

Grayson Dennis

The processing power of single-board computers (i.e., Arduino, Raspberry Pi, Jettison, etc.) allows for high throughput real-time data processing. Python is the primary language used within Raspberry Pi due to its versatility, usability, and extensive libraries, which grant the ability to easily access and manipulate incoming and outgoing data. Using the integrated GPIO pinout on a Raspberry Pi, 4-20 mA data can be acquired for any valves and instruments producing or acquiring this signal. Armed with the hardware and software to perform I/O, a heated continuous stirred tank experiment is controlled using <$400 worth of electronics. Students gain experience and understanding in process controls using control valves and gathering data which can increase their level of learning within the course. This presentation serves as a basis for others who want to teach students Python while also incorporating process controls, data acquisition, and control system integration.

379h Hands-on Microfluidic Glucose Sensor Module for the Enhancement of Engineering Courses

Bernard Van Wie, David B. Thiessen, Prashanta Dutta, Olusola Adesope, Riley Fosbre

Riley Fosbre,1 David B. Thiessen,1 Prashanta Dutta,2, Olusola Adesope3 and Bernard Van Wie,1, (1) Gene and Linda Voiland School of Chemical Engineering and Bioengineering, Washington State University, Pullman, WA, (2) School of Mechanical and Materials Engineering, Washington State University, Pullman, WA, (3) Educational Psychology, Washington State University, Pullman, WA

This project focuses on teaching microtechnology and microfluidics topics to the next generation of engineers by increasing the number of interactive experiences. Microfluidics deals with the processing of micro-scale quantities of fluids through micron-sized channels, allowing for the rapid emergence of breakthrough technologies meant for experimentation and analysis, such as point-of-care diagnostics and multiple clinical applications. These technologies have radically impacted chemical and biological analyses, allowing for improved understanding of cellular biology, disease diagnosis and progression, as well as fluid mechanics, while producing economical and disposable micro-devices due to advancement in soft lithography techniques. Unfortunately, robust teaching of these concepts has been challenging due to the topic's interdisciplinary nature.

Microfluidics encompasses several scientific fundamentals, including materials science, chemistry, physics, and biology, requiring training in all of them to take full advantage of multiple concepts. Currently, a significant gap exists in engineering curricula due to the engineering students' lack of robust understanding of microfluidics and its core principle. The long-term goal of this project is to address this gap with a low cost, hands-on educational series of modules, meant to advance knowledge of this complex topic while enhancing student engagement. In the short term, this means designing and implementing micro-devices which display the phenomena in question as well as accompanying educational material for use in chemical (ChE), mechanical (ME), and biomedical (BME) engineering courses at WSU (Washington State University) and WVU (West Virginia University). In addition, we seek to conduct ongoing assessments of this approach on student learning and engagement. An example of one of these modules currently in development is a glucose enzymatic analyzer, a micro-device which relies on microfluidic principles and a common compound analyzed in the medical field, glucose, to cover several scientific topics at once, including microfluidic mixing, reaction kinetics, and absorption spectroscopy based on Beer's law.

Glucose is commonly measured in blood chemistries because it is a key indicator for many medical conditions, such as diabetes and other metabolic diseases. The concentration of glucose in blood can be measured via a reaction in which glucose is first oxidized into gluconolactone and peroxide, the latter of which is then detected by reacting with 4-amino antipyrine in the presence of the enzyme peroxidase, which yields a final violet-red dye product, quinonimine, that absorbs over a characteristic wavelength range as an indicator of the concentration of glucose.

To that end, the experimental set up is simple. Students will inject reagent and sample into a Y-shaped chip with an on-off valve at the intersection. From there, the sample and reagent will be mixed within a micro-channel via an oscillating electric field. This leads to a color change reaction with a faint pink for low concentration, deep red for high. From there, the sample will pass into an opaque black box with a clear slit on the front side and a green plastic sheet on the other. This allows green light to pass through the sample when backlit. The green light that passes through the slit can be measured using a smartphone app, and with Beer's law, students can determine the concentration of glucose in the sample based off the measured intensity. Student comprehension of the concepts related to the experiment will be examined using accompanying worksheets, where the reaction kinetics and scientific principles can be tied in with the course. This experiment would work well with biomedical and reaction engineering courses.

After completing and testing the initial design, modules are planned for implementation by ChE, BE, and ME faculty at WSU and WVU using the modules, guided worksheets and accompanying assessment materials to assess learning of microfluidic concepts in relation to existing curricula. We will manufacture using simple and inexpensive 3D printing technology for rapid prototyping, allowing for fabrication of devices tailored to the unique needs of the intended course or experiment desired by the instructor.

We will present details on design construction and building of guided inquiry exercises and concept and motivational assessment instruments. We will also outline content for conducting surveys and interviews with instructors. This will provide the necessary feedback to ensure the integration of this additional curricula is well implemented by instructors.

The long-term goal of this program of research is the propagation and widespread adoption of new, more effective pedagogical tools and cutting-edge engineering technologies for teaching these complex topics - microtechnology and microfluidics.

379i Enabling Hands-on, Multi-Tiered Training, from High School to Industry Levels, on Batch to Flow Chemistry Process Transition

Fatou Baka Diop, Ashli Silvera, Shalom Njau, Fairuz Chowdhury, Andrew R Teixeira

Converting existing batch systems to flow requires knowledge on design and scale-up of bench scale experiments. In pharmaceutical processes, fast transport properties in flow microreactors can be critical for safe high-temperature and pressure reactions, or high-purity processes with enhanced mixing. Flow chemistry training is highly demanded in the pharmaceutical workforce as many companies seek to harness its benefits in product quality, process efficiency and safety. We have developed a laboratory training curriculum on the fundamentals of batch to flow process conversion with experiments and case studies relevant to small-molecule manufacturing. We hereby outline our unique multi-tiered approach designed to target different levels of chemistry and chemical engineering knowledge, from high school to industry. We achieved the flexibility and adaptability by layering high school level chemistry concepts, fundamental mass transfer principles, and real-world complexity with continuous-flow drug synthesis techniques, allowing to adjust the rigor while keeping core concepts and motivating objectives consistent. In this work, we 1) developed a batch to flow case study and lab experiment on the liquid-liquid extraction of salicylic acid, 2) defined learning objectives and calculations at tiered technical levels, and 3) deployed to the respective populations. The training was first deployed to a multilevel mentorship group comprising two high school students who focused on making a calibration curve for measuring salicylic concentration, one undergraduate student and one graduate student who worked on batch to flow extractions and explaining the underlying principles to the less experienced members of the group. The training program was also offered in a graduate reactor design course and a workforce-focused course in partnership with the RAPID Manufacturing USA Institute, providing hands-on exposure and advanced training on flow chemistry fundamentals to a total of 85 people.

379j Integrating a Vitamin C Iodine Clock Experiment into Undergraduate Kinetics

Alex Bertuccio

One of the challenges when teaching core chemical engineering courses is having a hands-on component as part of the course to strengthen student understanding of key concepts. An added benefit of having integrated lab experiments is that students can develop their own experimental plan to varying levels.

Here an iodine clock experiment was integrated into the junior undergraduate kinetics course to help students with determining rates, rate orders and activation energy. Students were given several pre-lab readings and a pre-lab homework assignment to motivate students to review the readings and to determine how to analyze potential data from the experiment. Students had a reference YouTube video (made by one of the professors) as a visual reference to aid in student understanding prior to the experiment. Students were given some instructions and suggested experiments to test and they completed a 3-hour experiment. Students obtained rates, rate orders and activation energy for their system and submitted a 3-page maximum report summarizing their findings. This experiment provided students with a hands-on experience, basic experimental design skills and technical writing practice.

379k Best Practices Using Low-Cost Desktop Learning Modules in Thermal-Transport Courses

Jacqueline Burgher Gartner, Bernard Van Wie, David B. Thiessen, Prashanta Dutta, Olusola Adesope

Over the past five years Low-Cost Desktop Learning Modules (LC-DLMs) have been nationally disseminated for implementation in thermal-transport classes. Faculty participants have used four miniaturized fluid mechanics and heat transfer modules in their undergraduate chemical and mechanical engineering courses collecting data on student cognitive gains and their motivational changes after implementation. The fluid mechanics hydraulic loss and double pipe modules have consistently had higher cognitive gains than the two heat transfer double pipe and shell and tube heat exchangers. All four modules have had positive motivational gains for students. This presentation highlights the best practices of implementations we have learned over the five years, discussing how to structure a classroom implementation, how to write and align supplemental materials like worksheets and assessments, and the features of the LC-DLMs that students identify as increasing their motivation.

379m Comparative Analysis of the Graduate Student Learning Experience for Native and Non-Native Students in the USA and Denmark

Heather LeClerc, Andrew R Teixeira

With the rise in graduate student enrollment in the United States comes further discussion of the most effective teaching methods and course design. The traditional graduate student in America is required to enroll and pass 30 credit hours of coursework, typically in semester format meeting 1-2 times per week for up to 16 weeks. This model often causes graduate students to feel that they are losing precious time that could be spent focusing on research and makes it difficult to schedule time-sensitive experiments around classes. In Denmark, however, the majority of PhD-level courses are intensive 1 week courses designed to minimize impacts to research productivity and typically last 5-8 hours per day for 3-5 days. In this format, students obtain both lecture-style information as well as hands on demonstrations and assessments.

The present study aims to analyze the differences between PhD-level coursework and subsequent learning objectives between the US and Denmark to make suggestions on how to improve the PhD experience in the United States. To this end, students studying engineering in both the US and Denmark were surveyed on their opinions of course offerings, objectives, and time commitments. Subjects were also asked whether or not they were native to the country in which their PhD studies were taking place. Survey results were then used to perform a comparative analysis and make recommendations on ways to improve PhD-level coursework in the United States.

379o Immunoengineering As an Elective for Senior Engineering Pre-Med Students

Tanyel Kiziltepe

One of the challenges in a Chemical Engineering program is the development of a pre-med curriculum that must integrate well into the traditional chemical engineering curricula while still preparing the pre-med students with the necessary foundation for successful outcomes in medical schools. Most students who choose pre-med track in a chemical engineering department are 1) either uncertain about pursuing a career between chemical engineering or medical field, and would like to learn more before committing to one, 2) they enjoy engineering approaches over basic sciences, such as biology, and believe that the problem solving skills and technology savviness developed in an engineering department will differentiate them among their peers in the medical school environment, or 3) would like to pursue an MD/PhD graduate program where they can simultaneously contribute to society on the bench side as PhD engineers as well as on the bed side as clinicians. Developing such a training program must take considerations that go well beyond the completion of bachelor's studies for a chemical engineering degree while simultaneously meeting all criteria required for preparation and acceptance into medical schools. This challenge becomes most apparent during the identification and development of senior level electives that can fulfill the tech elective criteria of the engineering department while providing students with the necessary foundation for medical schools. Here, the strategy to develop a course on Immunoengineering will be described. The course stretches over several multidisciplinary topics including: 1) the principles covered in an introductory immunology course, 2) dives deep into literature to discuss cutting edge research pursued by chemical and biological engineers in the field of immunology.

Immunoengineering is offered as a one semester senior/graduate-level course at Notre Dame, so that graduate students in addition to pre-med students who find it complementary to their PhD studies can also benefit from it.. As for the prerequisites, typically, one basic biology course with an emphasis on cellular systems is adequate, however, an additional biological course such as cell biology, biochemistry, or any bioengineering-based elective is recommended. The course is divided into 4 major modules where each covers a critical fundamental component of immunology followed by a relevant literature example of how engineers approach immunology in designing solutions to issues relating to human health. The assessment of the students involves homework assignments, exams, literature presentations, and a final term project where the students work on teams on a topic of their interest.

This talk outlines the approach used to develop an Immunoengineering course in a Chemical and Biomolecular Department which meets the tech elective criteria while at the same time providing the pre-med engineering students with the foundation in immunology to help them succeed in medical schools. This class has been proven to be very beneficial for not only the pre-med chemical students, but also for students interested in pursuing a career in pharmaceutical industry, medical device companies, medical or pharma arm of consulting firms, and also students planning of pursuing a PhD in a Biological Engineering related field. In the past 4 years this class has been offered, it has received very favorable feedback particularly from the recent graduates in graduate or medical schools who wanted to make a point in reaching out to us to let us know about their appreciation as the training they received significantly benefited them during their ongoing studies.

379p Inquiry-Based Learning for Students in the Senior Unit Operations Course with a Hydroponic Plant Growth Experiment

Jon Friedman, Jacob Lakomy, John Kopas, Ajay Patel, Alex Bertuccio

In 2011, the concept of a Water-Energy-Food nexus was first used and identifies that water, energy and food are interconnected. As the human population grows there will be an increasing demand of food and energy needed which puts a major strain on water resources as there is a finite amount of freshwater that can be used. The National Academies has highlighted that this is a new direction that chemical engineering should pursue.[1] One of the methods to increase food supplies to begin addressing the Water-Energy-Food nexus challenge is developing hydroponic- and aquatic-based technologies.[1]

As part of senior unit operations, one group of students developed their own experiments to investigate operating conditions for a hydroponic system. In the second semester of the two-semester senior unit operations course, seniors were tasked with choosing a topic, developing their own experimental plan, identifying variables to change and quantifying the effects of different conditions. Several variables tested for the hydroponic system were seed density, water flow rate and nutrient concentration and the growth of microgreens was monitored. Experiments were completed over several 2-week growth cycles and every cycle had 12 conditions tested in triplicate. The mass of drained growth pads was measured every other day and plant height and dry mass were measured at the conclusion of each growth cycle. Experiments demonstrated variations in plant growth based on the different growing conditions. Concurrently, students appreciated the ability to create their own experiment and investigate a topic of interest to them. The hydroponic growth experiment presents a new lab experiment that can be integrated into the chemical engineering curriculum and a means to expose chemical engineers to a new topic relevant to the field.

[1] National Academies of Sciences, Engineering, and Medicine. 2022. New Directions for Chemical Engineering. Washington, DC: The National Academies Press. https://doi.org/10.17226/26342.

379q Learnings from the Design, Commissioning and Operation of a Pilot-Scale Distillation column for the Unit Operations Laboratory at the University of Kansas

Felipe Anaya, Mark B. Shiflett

Construction has been completed for a new state-of-the-art pilot scale distillation column for the unit operations laboratory (UOL) at the University of Kansas (KU). The column, made of stainless steel, features 20 conventional sieve trays, total heigh of 20 ft, and diameter of 6.3 in. The addition of this equipment is part of an ongoing effort to update and modernize the UOL facilities at KU with the main goal of improving student's experience and learning. The unit operations laboratory course at KU is designed as a series of two courses taken during the Fall and Spring of the senior year. In the Fall, students carry out experiments on topics like thermodynamics, kinetics, fluids, and other related topics to their junior year coursework. In the Spring, they focus on unit operations experiments such as absorption or distillation and other topics from their senior year curriculum such as process dynamics and control.

The laboratory space has a high-bay and low-bay areas to house pilot-scale and bench-scale experiments, respectively. The new pilot-scale distillation column, located in the high-bay, is highly automated with variable frequency pumps, pneumatic valves, solenoid valves and many instruments such as RTDs, flowmeters, and pressure transducers. All the electronic I/O signals are read/written by a programmable logic controller (PLC) system and displayed on a computer running LabVIEW. The students operate the column using LabVIEW, which in turns operates the column through the PLC system. A dedicated control room houses the PLC and the industry-like workstations with 27-in dual monitor computers from which student operate the column. Additionally, tablets are wirelessly connected to the PLC to operate distillation column while doing walkthroughs in the high-bay area. Very few operational steps during start up require to be physically present in the high bay area allowing to operate the equipment from the control room. In this presentation, we will share some of our learnings from the design, commissioning, and construction of this experiment. Additionally, we will share our experience operating the column with students along with student's feedback from their learning experience. Finally, we will discuss planned strategies to optimize the utilization of column features to enhance student learning.

379r Teaching Semiconductor Fabrication to Chemical Engineers

LiLu Funkenbusch, Sumant Patankar

At the University of Florida, we have a wide range of experiments in our teaching laboratory courses. Many of these are the more traditional concepts and unit operations that are common in teaching labs everywhere, such as fluid flow, distillation, and process control. One of our more unique experiments is semiconductor fabrication. Through a series of four lab periods, students perform one "iteration" of the semiconductor fabrication process. This provides an opportunity for them to learn more about an expanding engineering field as well as explore certain chemical engineering concepts further in an alternative application. At the graduate level, the modules are expanded to allow students to optimize process parameters. This provides students with a preview of challenges faced by process engineers and systematic DOE approaches.

In this talk, we will outline the order of these semiconductor experiments and which chemical engineering concepts they explore with each experiment. We will showcase our unique facilities and experiments while also providing a base model for others to explore less traditional areas of chemical engineering in their own teaching laboratories.

379s Technical Writing Rubric Design Based on Inter-Rater Reliability

Jennifer Brown, Stephanie Wettstein

Creating rubrics that multiple instructors used to grade technical writing from consistently is challenging. The purpose of this work was to evaluate the inter-rater reliability (IRR) of a rubric used to grade technical reports in a senior-level chemical engineering laboratory course that has multiple instructors that grade deliverables. The rubric consisted of fifteen constructs that provided students detailed guidance on instructor expectations with respect to the report sections, formatting, and technical writing aspects such as audience, context and purpose. Four student reports from previous years were scored using the rubric, and IRR was assessed using a two-way mixed, consistency, average-measures intra-class correlation (ICC) for each construct. Constructs were rated from poor to excellent based on their ICC and then, the instructors met as a group to discuss their scoring and reasoning. Revisions to the rubric were made and the IRR completed again. A key learning from this process was the importance of the instructor discussion around their reasoning for the scores and the importance of an 'instructor orientation' involving discussion and practice using the rubrics in the case of multiple instructors or a change in instructors. Additionally, the fewer constructs that instructors interpreted has having potential overlap (e.g., double-counting errors), the more reliable the grading. The developed rubric has the potential for broad applicability to engineering laboratory courses with heavy technical writing components and could be adapted for alternative styles of technical writing genre.

379t Improving Undergraduate Success through Effective Critical Thinking

Nathan Duran Ledezma, Twanelle Majors, Stephanie Wendt, Indranil Bhattacharya, Joseph Biernacki

Employers report that most of their first-time hires, usually recently graduated students, lack critical thinking skills and problem-solving dexterity. Tennessee Tech's *Undergraduate Success Through Effective Critical Thinking* (iUSE-CT) pilot study has two major goals regarding academic success and knowledge acquisition: 1) enforce first year freshmen chemical and electrical engineering students' retention rates, and 2) improve graduation rates. The methodology implemented consists of deliberately training students to identify and use critical thinking skills through individual activities as well as group tasks. Assessment is a combination of a retrospective Need for Cognition survey and pre- and post-treatment *Critical Thinking Assessment Test* (CAT). The CAT is a widely used product of a decades-long *National Science Foundation* (NSF) funded effort that assesses a well-defined range of CT skills using metacognition. The study design includes two cohorts of students divided into treatment and control groups. The design uses a two-level full-factorial matrix in which students are either included in the treatment or control groups as freshmen or sophomore students in required, critical-path courses in either chemical or electrical engineering. The courses are paired across disciplines at analogous points in the students' development and the CAT is administered at the beginning of freshman year, and again at the end of the first course (and in subsequent terms). Critical thinking training includes sessions on skills and fast-paced, in-class assignments and extended take-home work using situational open-ended frameworks. Most of the exercises are presented in a laboratory setting where students have a significantly longer class-time to accommodate extended discussion with their classmates and the instructors.

129b How We Teach: Chemical Engineering Electives

Laura Ford, Janie Brennan, Jennifer Cole, Kevin Dahm, David Silverstein, Stephen Thiel

The Fall 2022 survey by the AIChE Education Division Survey Committee is over chemical engineering electives. The survey will collect information about how many electives are required for chemical engineering undergraduate students. The specific focus is chemical engineering elective courses rather than engineering or science elective courses. We will survey the topics of the offered electives and compare to the results of previous surveys.

129a Process Safety across the Curriculum

Laura Ford, Christopher Barr, Janie Brennan, Tracy Carter, Kevin Dahm, Hunter Flodman, Luke Landherr, David Silverstein, Stephen Thiel, Bruce Vaughen, Troy Vogel

The AIChE Education Division Survey Committee is conducting a survey over process safety in the curriculum during spring 2023. Departments have responded to ABET requirements for process safety training in different ways, and these are not necessarily captured by the annual fall surveys over specific courses and the curriculum overall. Questions include where students receive process safety instruction (a single course, multiple courses, other?), what Bloom's level students are expected to achieve in process safety and management-of-risk outcomes, and usage of various AIChE and other safety resources. Data from AIChE about the popularity of specific SAChE modules will be presented in addition to the survey results so that the departments may gauge themselves against other universities and gain ideas for new safety training.

25a Perspective on Starting a Mentoring Program for First Year Ph.D. Students

Mariah Arral

Entering graduate school is a time of transition in many students' lives and often correlates with students moving to a new location and removal from their support networks. A study on engineering graduate student thriving found that students who persisted in their degree may only be surviving, rather than thriving. More research is needed to understand engineering graduate students thriving, but researchers have shown that mentorship positively affects academic and learning outcomes in higher education. Providing students with mentors when they join a new department is one strategy in supporting engineering students thriving. This presentation will detail the creation of a first-year mentoring program at Carnegie Mellon University in the Department of Chemical Engineering. The conceptualization and implementation for the first year of the program. This first-year mentoring program is completely graduate student run. The goal of this presentation is to provide attendees with a perspective on the creation of a first-year mentoring program and in hope that others will create one in their program if one does not exist. Furthermore, this presentation will detail the iteration that occurred over the past three years. At the end of this presentation, attendees will know 1) about the importance of mentorship during transitional time periods, 2) strategies for starting a first year mentoring program, and 3) retrospective analysis on changes that would have been made earlier.

25b Teaching Chem-E Journal Club: An Opportunity for Career Exploration and Student Support

Rebecca Harmon

Journal clubs are often used as a tool to increase exposure to a field of interest and develop a variety of skills like critical thinking and application of evidence-based practices. A monthly journal club was coordinated in a chemical engineering department that has traditionally not had formal conversations about teaching pedagogy or pathways to teaching in higher education. These meetings were open to undergraduate and graduate students and postdoctoral fellows while being facilitated by clinical-track faculty. A heterogeneous group was intended to allow for mentorship and exchange of experiences for early-career chemical engineers. Each monthly meeting was led by a different member who picked a research article and facilitated discussion.

This journal club focused on discussing engineering education research, particularly research applied to the chemical engineering classroom. In addition, a lens of equity and inclusion of historically excluded groups was added to consider the impact of evidence-based practices. This lens was chosen because of how interconnected engineering decisions are to the broader community and was intended to get early-career engineers thinking about their impact both in the classroom and through their work products.

Outcomes from this trial are mainly intangible through exposure to teaching techniques, exploration of teaching-track careers, and improved critical thinking of impact on belonging, equity, and inclusion. Feedback about the journal club and outcomes was collected using an initial survey, mid-term survey, and closing survey at the end of the academic year. As a result of the journal club and participation by current students and alumni, several ideas for positive change in the department include a syllabus addendum on student support and supplemental teaching assistant training for undergraduate and graduate students.

Journal club is a time-effective tool to discuss teaching methods and their impact on student support in chemical engineering for people who are interested in exploring a career path related to teaching. Future work is needed to study the motivation of attendees and how to encourage attendance by industry-focused department members.

25c Green Chemical Engineering Innovation: Turning Design Constraints into a Source of Creativity

Christopher V. H-H. Chen

There are many drivers that spur student interest in green and sustainable solutions–from addressing climate change and protecting the environment; to a feeling of moral or ethical duty. However, the conditions that create green innovation in industry differ from those which drive students to study the topic. Though ethical motivation may affect the behaviors of individual companies or people therein, firms are also driven by regulation, their stakeholders, and finances in their push towards sustainability. For chemical engineers, many of these green considerations focus on constraints: moving towards the removal of petrochemical feedstocks, limits on the types of emissions, designing to remove extreme process conditions, etc.

To help students move beyond seeing sustainability as an additional burden to the design process, we have developed a graduate-level green chemical engineering course that focuses on how to use the principles of green chemistry and engineering to create value–attracting students who may be driven by internal motivators, but providing them skills that help them become more valuable to future employers. Students are introduced to Entrepreneurial Mindset (EM) and Systematic Inventive Thinking (SIT) as lenses through which to undertake green chemical engineering. We use the KEEN 3C's framework–curiosity, connections, and creating value–to center potential impact within the course. This is not to strictly encourage students to create their own start-ups or to go into business per se, but instead to help them recognize that they have the ability recognize opportunity and realize impact as they are learning in this course.

Systematic Inventive Thinking was borne out of previous theories that find patterns in innovation, such as TRIZ (Russian abbreviation for Theory of Inventive Problem Solving). SIT reduces the 40 TRIZ patterns into five templates following three core principles. The first of these principles is the Closed World, which states that creativity comes from closure instead of thinking "outside of the box." For the case of this course, the Closed World are the principles of green chemistry and engineering, and the laws and regulations that bound chemical engineering design. The second is principle is Contradiction, where two demands for a product or process are often connected and are in opposition. SIT provides methods to disentangle contradictory design conditions to allow the contradiction to exist within a product. The last principle is Function Follows Form (FFF), the opposite of Form Follows Function. FFF states that potential, already created solutions are easier to connect to problems that are ill-defined than the reverse (this could be thought of as "solution-probleming" instead of problem solving). Simply put, FFF is asking what value a potential design could create, whether or not it is for the application it was originally intended. Students use these three principles and the five SIT templates–multiplication, division, subtraction, task unification, and attribute dependency–to understand previous green innovations and pitch their own at the end of the course.

This green chemical engineering course was designed to facilitate students to be able to do the following by the end:

1. Describe the tenets of green chemistry and their pertinence to chemical engineering practice.
2. Explain how Systematic Inventive Thinking (SIT) and green chemistry constrain the design process and can help improve innovation.
3. Apply green chemistry approaches to the design of a more sustainable product/process.

4. Evaluate the environmental, business, and social impacts of a chemical product or process.

Although EM was not an explicit learning goal for students in the course, the 3Cs pervade the course design as summarized in **Table 1.**

Grading *participation* encourages students to make connections between the course content with a focus on creating value. Forming *fixed learning teams* for the term (groups of 4-5) was done with the intention that students would be able to foster at least a small community of peers within the class that could form the basis of other professional interactions in the future. These groups were created by the instructors based on some shared affinity of the students (e.g., sector, career aspirations, or research interests), and students completed all group learning activities within these learning teams–both in and out of class. *Community Contributions* are graded as a means of recognizing individual student curiosity and encouraging them to share their curiosity with the class. These activities included readings that students individually found and shared back with the course through online discussion forums, and through different activities through which students leveraged their own goals for the course to contribute to greater classroom dialogue. *Homework* provided exercises to help students practice individual skills on way to the final class project. The *Green Design Project* required students to pitch a green chemical product and/or process, where students needed to draw on the technical content of the course, along with using the SIT templates to create such an innovation. This project ran for the final third of the course and was completed in their learning teams.

At the end of the course, students are asked to report on their learning in relation to the KEEN 3C's, along with their general learning experience in the course. We are especially interested in how the students' experience in this course compares to other courses with sustainability themes or content in engineering. As this course is being taught for the first time this term, we do not have current data to report on for these research goals. However, students have already commented on how the course takes a different approach in discussing innovation and engineering design in generally positive ways. SIT has provided students a means through which they can follow to start generating ideas–though students have needed practice in being able to connect these new ideas to potential applications. EM and SIT together (and specifically the Function Follows Form principle) has allowed for there to be greater emphasis on the "why" of innovation over the "what" or technical "how" that is common in engineering education. This prioritizing of whether an idea creates value and for whom prior to diving into the details of making an invention work, which has been positively received by students thus far, is an approach the instructors are interested in carrying over into other engineering courses. This emphasis also distinguishes this course amongst others focused on sustainability and green technology at the engineering school.

By the Fall, we plan to report more formally on the students' experiences through the aforementioned approaches along with an analysis of student work.

25d Retrospective Analysis of the Effect of an Online "Bridging" Course Sequence on Student Success in Graduate Studies

Matthew Cooper, Lisa Bullard, Aaron Frye

Since 2018 NC State's Department of Chemical and Biomolecular Engineering (CBE) has offered a self-paced online course sequence consisting of two courses, three credit hours each, for students who have a BS degree in a non-ChE science or engineering field. This two-course sequence serves as a "bridging course" and teaches core ChE concepts to students, covering material including material and energy balances, thermodynamics, transport phenomena, and reactor design. The goal of this work is to determine the impact, whether positive or negative, of bridging course sequence completion on the success of students without undergraduate ChE degrees in CBE graduate studies; in other words, we aim to determine if the bridging course sequence is more effective than ad-hoc measures (e.g. having students take designated undergraduate classes) at helping students achieve success in ChE graduate study. Here we present assessment of whether an experimental group of 77 students who have taken the bridging course since 2018: (1) performed better in core CBE graduate courses (Process Modeling, Thermodynamics, Transport Phenomenon, and Chemical Reaction Engineering) and (2) more frequently earned a graduate degree compared to a control group of 155 students who were admitted to the graduate program since 2009 without a BS ChE degree and did not take the bridging course. We also provide an analysis of only bridging course students which contrasts student GPA across the bridging course sequence against GPA in their later core graduate ChE courses to explore whether success in the bridging courses portends success in specific core graduate courses of the curriculum. The results of this work will support advising of prospective graduate students without ChE undergraduate degrees as well as identifying potential areas for improvement in bridging course content.

528a Development of Elective Courses for Process Industries Certificate As Recommended By the Industrial Advisory Board

Matthew Alexander

The majority of our chemical engineering graduates find employment at refineries, chemical plants, or industry-related design or services firms in our south Texas region after they graduate. The industrial advisory board for our department, which includes engineers from these same industries, indicated in their 2021 meeting a desire for our chemical engineering graduates to have exposure to some course that relate more directly to activities the new graduates would experience in industry. Based upon this recommendation, our chemical engineering department developed a new certificate (12 course hours) program, which was given the title the Chemical Process Industries certificate. The new certificate program was proposed and approved in Fall 2021, with new elective courses in process safety, sustainability, environmental control of chemical processes, and air pollution control constituting this certificate. Additionally, a course was added in introduction to chemical engineering for non-majors, so that engineering students from outside the major could engage in the certificate program as well.

The department faculty were skeptical at first that there would be sufficient interest amongst our students in discipline-specific elective courses. There are a couple of reasons for this mindset, first because the curriculum for the chemical engineering BS degree requires 130 hours, and second because there have been minimal offerings of elective courses at the undergraduate level in our discipline for a number of years. However, this faculty member has recognized there is some demand for upper-level electives in chemical engineering because some seniors end up needing only 11 hours (12 hours required for full time student status) of required courses for both of their senior year semesters, and because some students look for minors to complete beyond the chemistry and math minors that all chemical engineering majors can readily attain with their BS degree. The first of these elective courses for the new certificate program were offered in the 2022-2023 academic year. In the first and second course offerings (safety course and sustainability course), the enrollment has been 15 and 17 students, respectively, thus confirming that the interest amongst our undergraduate students. The student feedback on the initial offering of these electives has been very positive.

528b Exploring Perceptions and Attitudes of Chemical Engineering Students Towards Data Science: Implications for Curriculum Integration and Industry Preparedness

Betul Bilgin, Daisaq Cyrilla, Naomi Groza, Ruben Lopez

Exposure to and expertise in data science becomes more important as the chemical engineering (CHE) industry incorporates novel analytics and large-scale computational techniques. With data science becoming increasingly important in the chemical engineering industry, it is crucial to understand how students view this field and its integration into their curriculum.

This study surveyed a sample of chemical engineering students from various class levels to assess their understanding of data science and their attitudes towards incorporating it into their coursework and future careers. The survey instrument was developed by our collaborators [1] and is composed of four constructs: interest, career, value, and self-efficacy. The survey included questions on students' prior exposure to data science concepts, their perceived importance of data science in chemical engineering, and their willingness to engage with data science tools and techniques. Senior and sophomore CHE students are surveyed at the beginning and end of the semester. Moreover, focus groups are interviewed and their responses are analyzed.

The results of the study revealed that while most students had limited exposure to data science in their coursework, they recognized its importance in the chemical engineering industry. The majority of students were interested in learning more about data science and were open to incorporating it into their coursework and future careers.

The study also identified barriers to incorporating data science into the chemical engineering curriculum, including the need for additional resources and training for both students and faculty. The findings of this study provide valuable insights for chemical engineering educators and industry professionals seeking to integrate data science into their curriculum and practices.

This talk will provide a detailed explanation of how the surveys and interviews were developed and analyzed, data distribution and findings. It will highlight the need for further research to better understand how to effectively incorporate data science into the chemical engineering curriculum, as well as the importance of preparing students to work in an increasingly data-driven industry.

1. Bonfert-Taylor, Petra, Laura Ray, Scott Pauls, Lorie Loeb, Logan Sankey, James Busch, and Taylor Hickey. 'Infusing Data Science into the Undergraduate STEM Curriculum.' In *2022 ASEE Annual Conference & Exposition*. 2022.

528c Scale Down Studies to Address Chemical Production and Student Development: Case Study of a Partnership with a Chemical Manufacturer and Technical Elective Design.

Robert Beitle, Tammy M. Rechtin, Dearl Peachee, Mitchell Cole, Tony Maza

This case study illustrates the synergy that is possible when the desire to synthesize a chemical intermediate by an industry partner combines with the desire to develop relevant, hands-on educational experiences for chemical engineering students. A technical elective was created for an undergraduate / graduate offering that combined the topics of process scale down, the writing of standard operating procedures (SOPs), and hazard mitigation to take students through the methodology needed to critically evaluate and test a synthesis route. Care was taken during course design to minimize difficulties with intellectual property considerations, chemical safety, departures from SOPs, and other pitfalls common to 'new' class experiences. Course exercises addressed two principal ABET outcomes: outcome 6, *an ability to develop and conduct appropriate experimentation, analyze and interpret data, and use engineering judgment to draw conclusions*; and outcome 2, *an ability to apply engineering design to produce solutions that meet specified needs with consideration of public health, safety, and welfare, as well as global, cultural, social, environmental, and economic factors.* With guidance, students designed experiments for a 20-gallon batch system installed at the university, went through the approval processes to test a Kolbe-Schmitt reaction, refined SOPs, worked with visiting engineers from the company to evaluate process chemistry, and participated in discussions to translate results to a proposed 1,500 gallon system. Additional process improvements are anticipated for future offerings of the course, designed in a manner to minimize any difficulty with intellectual property considerations.

528d The Two-Semester Senior Capstone Design Course Sequence at UMass Lowell – Preparing Students for Engineering Careers in Industry

Eric L. Maase, Glenn Dissinger

The senior capstone design course is a bedrock core course required by ABET in all engineering disciplines at all universities in the United States and probably throughout most of the world. Ideally, the implementation of a comprehensive capstone design project provides the soon-to-be chemical engineering graduate a chance to leverage the skills and knowledge learned over their undergraduate career in a group project that prepares them for the exciting challenges they will face as a practicing chemical engineer, be it in further academic graduate studies, or more likely, in a new position within industry.

Here at UMass Lowell, we have developed a comprehensive two-semester course sequence that incorporates many of the skills and challenges our graduates will face in their careers. Components of our program include:

- Six week intensive Engineering Economics segment in the Fall Course where students are introduced to the economics of business and project evaluations, including the Time Value of Money, Cash-Flow Analysis, Rates of Return Metrics (PW, FW, MARR, IRR, etc.) Depreciation, and Pro-Formas.
- Three group projects (3-4 students per group) in the Fall Semester:
    - Fundamentals Project – a four-week deep dive project into a particular piece of equipment (compressors, membranes, filters, relief valves, etc.) or process (LL Extraction, Crystallization, Pollution Abatement), including design fundamentals, selection criteria and costing.
    - Feasibility Project – a four-week study on the feasibility (technical, regulatory and market) for a potential new processing plant to manufacture a chemical within the United States.
    - Design Project – a three-week "simple" design project to introduce the design process.
- A full comprehensive capstone design project during the Spring semester done by "companies" of 6-7 students for a new greenfield plant to be designed and built within the continental United States. Each student is assigned two specific roles: one technical (e.g., Process Design Equipment – Separations, Material Transport Engineer – Liquids, Process Control, etc.) and one non-technical (e.g., Chief Financial Officer, Marketing, Human Resources Officer, Project Manager). The project deliverables include:
    - A comprehensive FEED package (PFDs, P&IDs, Piping, Facility Layouts, Equipment Specifications, etc.)
    - A 20-year economic analysis of the project of the economic feasibility of the project, including proposed financing for construction and start-up of the facility and all capital and operating costs (raw materials, personnel salaries and benefits, utilities)
    - A "go-to market" strategy and analysis on how the company will succeed as a new venture.
    - Oral presentation (e.g., "pitch") of their project to outside interests and potential investors.

A more detailed description of the many elements and evolution of this capstone design sequence will be covered, including plans for how it may evolve in the future.

528e Developing and Deploying Augmented and Mixed Reality Platforms for Chemical Engineering Labs and Workforce Development

Jacob Crislip, Fatou Baka Diop, Robert Dempski, Andrew R Teixeira

Augmented reality (AR) takes inputs from a real environment (sensors and cameras), and projects a holographic, interactive virtual environment on top of it in real time to create a 'mixed reality' interface. In this work, we explore four implementations of mixed reality for chemical engineering education:

- Synchronous remote/hybrid unit operations laboratory
- Asynchronous laboratory unit operation training
- Workforce development on batch-to-continuous flow

The first two were conducted during the senior Unit Operations course at Worcester Polytechnic institute, and demonstrated the power and pitfalls of AR. Microsoft Hololens 2 was used to superimpose tutorials in lab for individual unit operations. The later was a custom-application developed with ACS Green Chemistry Institute and AIChE RAPID Manufacturing Institute for a batch-to-continuous processing workforce development course. It quantitatively demonstrates the power of AR to engage and promote learning for laboratory courses, specifically with a tutorial on how to construct a packed bed reactor.

414e Employing Undergraduates to Develop Programming and Automation Modules for the Undergraduate Curriculum: A Two-Fer Scheme

Maureen Tang

In the 2020-21 academic year, the COVID-19 pandemic highlighted a) the need for agility in teaching and research; b) the desirability of computing skills for graduating students and c) the amenability of computational research to remote work. Moreover, although computing skills are increasingly paramount to success in an increasingly automated society, chemical engineering curricula are already heavy with coursework and leave little room for additional required content. Any application of modern automation tools must be incorporated into a traditional framework of thermodynamics, transport, and kinetics.

At the same time, the economic fallout of the pandemic generated a major problem for undergraduate co-op employment at Drexel University. Experiential education through co-op is the most distinguishing aspect of Drexel's educational mission. Drexel students typically complete three six-month industrial co-ops throughout their bachelor's. In 2018-2019, 98% of engineering students were employed as co-ops. In 2021, the financial fallout of the pandemic caused many employers to cancel their co-op programs, such that only 60% of Chemical and Biological Engineering students were able find positions for the April – September co-op cycle.

To address these acute and long-term needs in 2021, the CBE department developed a program in which stranded co-op students are hired to develop instructional modules for the undergraduate curriculum. These modules use hobbyist electronics, image analysis, and other low-cost methods to measure thermodynamic, transport, and kinetic properties with an emphasis on programming and automation. The experience provides important training opportunities to students who would otherwise be unable to find technical employment for the six-month period, with a deliverable hands-on device or interactive simulation that can readily be incorporated into existing lesson plans in the CBE curriculum. This talk will share the results of offering the program, including best practices, lessons learned, and suggestions for how similar programs could be implemented in other situations.

414b A Short Course Introducing Students to Computational Thermodynamics

Luc Paoli, Pierre Walker, Pavan Inguva, Paul Yew, Andrew J. Haslam

Thermodynamics stands as one of the fundamental pillars of a chemical engineer's educational background. Despite its theoretical underpinnings, computational elements have become increasingly important for contemporary engineering applications. Regrettably, most traditional undergraduate thermodynamics courses provide little exposure to practical computational applications, leaving students ill-equipped to engage with the complex thermodynamic models used in industry. However, the rise of open-source codes has created a unique opportunity for students to bridge this gap. To address this need, we have developed a comprehensive short course that leverages these tools to enhance students' computational thermodynamics knowledge base. The course is structured into three distinct parts. The initial segment provides a robust foundation in both numerical methods and fundamental thermodynamics principles, ensuring that all participants possess a shared knowledge base. The second part introduces a range of thermodynamic models, including activity coefficient models, cubic equations of state, and SAFT equations, and provides practical examples of their implementation. In the final segment, students are tasked with implementing their computational thermodynamics methods, covering pressure solvers, flash problems, and chemical stability analysis. In a pilot study involving both undergraduate and graduate students, we found that participants found the course engaging and highly relevant to their chemical engineering education. All course materials are openly available on GitHub, with future expansions planned to provide even more extensive resources. By providing students with a strong foundation in practical computational thermodynamics, we hope to empower them to engage with the ever-growing complexity of modern industrial applications.

414c Chemical Reactor Sequence Optimization Using Live Scripts in Matlab

Aycan Hacioglu, Walaa Abdallah

Reaction kinetics and reactor design courses are at the heart of the chemical engineering curriculum. In these courses, students work with algebraic equations, differential equations, numerical integration, curve fitting, and parameter estimation. Such tasks require computational tools such as MATLAB, a high-level technical computing language used in many chemical engineering programs. Computational tasks of reaction kinetics can be performed easily in MATLAB using interactive, executable notebooks – Live Scripts.

Inspired by Fogler's CSTR and plug-flow reactor sequencing problem in the "Elements of Chemical Reaction Engineering" textbook, a reactor sequence optimization question was generated that involves computationally exploring various quantities and sequences of CSTRs and PFRs with the goal of minimizing total reactor volume (thereby reducing cost) while still achieving a desired conversion. We present this problem, and its solution, as a Live Script where students can modify the input code for their particular problem and explore different variants of this problem while learning reactor design and optimization principles in an interactive way.

414d Design of Core Courses on Computing and Data Analytics in Our Chemical Engineering Program

Richard Davis, Weiguo Xie

Skills for computing and data analytics are in high industrial demand and becoming critical to our students' future career success. As our industrial advisor boards recommend, we continue enhancing our students' learning of computational methods, modeling, simulation of chemical engineering processes, and data-driven process optimization. It is crucial in a systematic way to design core courses on computing and data analytics in our chemical engineering program to bridge the skills gap.

We first designed the CHE 3031 Computational Methods course in 2010 as recommended by our industrial advisor boards. The original topics include Excel for engineering analysis, Pivot Tables and Charts, VBA programming, Linear equations, Derivative approximations, Roots of nonlinear equations, Optimization, Uncertainty, Least-square regression, Interpolation, Integrals, Initial-value problems and ODEs, Boundary-value problems, and PDEs. We added Python for scientific computing and data analytics in the course 2019 as recommended by our industrial advisor boards.

Secondly, the CHE 3032 Process Modeling course was designed as a computer lab-based course to train students in using commercial simulation and modeling tools such as Aspen HYSYS and SolidWorks to prepare them for our unit operations and design courses.

Thirdly, CHE5031 Chemical Engineering Analysis was designed to teach topics including Introduction to modeling in chemical engineering, MATLAB skill preparation, Model fitting, spline functions, Optimization toolbox, Enhanced optimization, Global optimization, Optimal experimental design, Data statistics and analytics, Complex equation systems, Process integration, and optimization. MATLAB was selected as the computational language for this course in 2020.

Computational Methods and Process Modeling are required core courses in our undergraduate program. Our graduate program requires the Analysis course, and undergraduate students may take this course to fulfill a technical elective requirement. As shown in Figure 1, these courses aim to train our chemical engineering students to higher proficiency with modeling and simulation methods and computer software tools. It can be seen from student evaluations that they enjoyed those courses and realize the importance of computing and data analytics through those courses. These courses also enable them to use the "data science" techniques, skills, and modern engineering tools necessary for engineering practice.

- CHE 3031 - Computational Methods in Chemical Engineering
- CHE 3032 - Chemical Process Modeling and Simulation
- CHE 5031 - Chemical Engineering Analysis

Figure 1: Our three courses related to computing and data analytics

414a Machine Learning for Engineering Course in Matlab and Python

John Hedengren, Andrew Fry, Sean Last, Jacob Burrell, Nathan Phillips, Peter Cook, Hannah Hanson, Sion Jung, Spencer Larson, Andrew Crop, Krystian Perez, Aycan Hacioglu, Jianghao Wang, Jon Loftin

A surge in the chemical engineering industry interest in machine learning has generated a high demand for students who possess expertise in this domain. Many chemical engineering departments are looking for ways to offer elective machine learning courses or integrate machine learning into existing chemical engineering courses. Chemical engineering instructors encounter a challenge in curating readily available curriculum materials that cater to the engineering domain when offering a machine learning course. Other aspects to consider are the familiarity of instructors with programming, student programming knowledge, and choice of software. MATLAB and Python are among the popular tools for machine learning. MATLAB's app-based workflows, self-paced online training courses, and Live Scripts lower the learning barriers for students and make it easy to introduce machine learning.

In this talk, we cover how the "Machine Learning for Engineers" course was translated from Python into MATLAB through the collaboration of MathWorks and Brigham Young University as part of the capstone project course and showcase translated modules. During this project, students were exposed to both programming languages, received help from domain and AI experts to apply machine learning to chemical engineering applications, and explored the use of ChatGPT for code translations. MATLAB modules developed in this project are freely available at https://github.com/APMonitor/mds on GitHub and File Exchange. These modules can be used as instructor-led course material or self-paced learning resources and they contain case studies related to additive manufacturing, concrete strength, lithium-ion batteries, steel plate defects, polymers, and thermophysical properties which are relevant to chemical engineers. Modules also cover fundamental machine learning concepts. By making these course modules available in MATLAB and Python, we aim to fill the gap in machine learning teaching resources in a platform of the instructor's choice.

414f Chatgmp: Introducing Virtual Tutors in Chemical Engineering Education

Fiammetta Caccavale, Carina L. Gargalo, Krist V. Gernaey, Ulrich Krühne

The overwhelming attention that ChatGPT, based on the GPT-3 model series [1] and developed by OpenAI, has gathered recently, is just one of many signs that the "Artificial Intelligence Revolution" is upon us. Also in education, artificial intelligence (AI) is rapidly becoming an integral part of digital platforms. It is in fact used across various applications, such as generating video captions in software like Microsoft Teams®, to develop smart content, grade assignments and personalize the learning of each student based on their individual profile. This is motivated by the fact that this technology has the potential to answer the needs of both educators and learners. For example, Lu et al. introduced RadarMath [2], an intelligent tutoring system to support personalized learning for math education. Another example of AI used in education can be provided by Carnegie Learning with 'MATHiaU' [3], a tool that uses cognitive science and AI methods to create personalized lesson plans and tutoring that adjust based on students' performance, and it also provides real-time feedback.

Among other benefits, AI provides the opportunity for educators to replace repetitive tasks with stimulant new experiences. This work aims to investigate the use of AI in this context, as a powerful tool to automate tasks and generate answers from previously collected data. The Department of Chemical and Biochemical Engineering at the Technical University of Denmark (DTU) offers a course in *Good Manufacturing Practice (GMP).* As part of the course, students are required to take part in an audit exercise, where they take on the role of an auditor investigating a company, represented by the teachers, about their good manufacturing practice. During the audit, students can ask questions regarding available documents and practices adopted by the company, which is necessary information to evaluate whether the company in question could be a valuable business partner or if detected non-conformities are too severe. The teachers' role is to give both accurate, in case the documents are available, or vague responses, leaving the students to reflect upon the company's behavior. After a few years of participating in the aforementioned audit exercise, the teachers have agreed to test the replacement of the physical teacher with an AI-powered virtual assistant. The hypothesis is that students will be more engaged by the gamified nature of the virtual assistant and will not be afraid to ask for any follow-up or further clarification of the given answers. Finally, the students will have the opportunity to practice more with the virtual tutor before the actual audit exercise, hopefully strengthening their learning.

This work presents the early stages of this experiment, aiming to develop ChatGMP, a digital audit tool to represent the fictional company and therefore replace the teachers. In practice, the virtual tutor consists of a pre-trained question-answering generative model (Transformer-based [4]), which is fine-tuned on domain-specific material. This ensures that the model contains both the general knowledge, given by the original pre-trained model, as well as the specific GMP material, derived from fine-tuning. In order to enable the students to interact with the virtual tutor, we build an interactive GUI.

The aim of this initiative is twofold: (i) to improve chemical engineering students' learning experience and increase engagement by providing stimulating material and gamification, and (ii) to allow substantial time saving for the teachers. A survey conducted shows that students seem to be interested in the option to have the possibility to perform the audit exercises with a virtual tutor. Moreover, quantitative interviews were conducted with a small number of volunteers to test the model and they reveal a positive perception of the generated answers by the students.

References

[1] Brown, T., Mann, B., Ryder, N., Subbiah, M., Kaplan, J. D., Dhariwal, P., ... & Amodei, D. (2020). Language models are few-shot learners. Advances in neural information processing systems, 33, 1877-1901.

[2] Lu, Y., Pian, Y., Chen, P., Meng, Q., & Cao, Y. (2021). RadarMath: An Intelligent Tutoring System for Math Education. *Proceedings of the AAAI Conference on Artificial Intelligence*, 35(18), 16087-16090. https://ojs.aaai.org/index.php/AAAI/article/view/18020.

[3] Carnegie Learning, MATHiaU, available at: https://www.carnegielearning.com/solutions/math/mathiau/, Accessed: 21-03-2023.

[4] Vaswani, A., Shazeer, N., Parmar, N., Uszkoreit, J., Jones, L., Gomez, A. N., ... & Polosukhin, I. (2017). Attention is all you need. Advances in neural information processing systems, 30.

414g Takeaways from Teaching Machine Learning

Jindal Shah

In recent years, data science and machine learning methodologies have been finding an increasing use in all science and engineering domains. Thus, there is a need to introduce these topics to the next-generation of scientists and engineers. In this presentation, recent efforts at developing a course focused on machine learning in the School of Chemical Engineering at Oklahoma State University will be described. The presentation will focus on the course content, mode of instruction, class engagement strategies and student feedback. Challenges of teaching such a class from the perspective of someone not trained in the subject will be discussed along with opportunities that exist as the course evolves.

414h Transition of Computational Software in ChE over the Past Four Decades

Aydin Sunol

This presentation will focus on use of flowsheet simulator such as Aspen plus, mathematical packages such as a matlab and languages such as pyhton in Chemical Engineering programs in the past 40 years. This will reflect experience in my home institution enriched by sabbatical experiences at US and abroad as well as shared practices in the peer community/organizations. Teaching capstone design series as well as systems/math related courses allow holistic perpective. The curricular implications, faculty buy-in and student perfonmance will be discussed as well as their evolution over time.

414i Experience with Using Symbolic Algebra and Smart Computing in Chemical Engineering Courses

Satish Parulekar

Experience in using symbolic algebra and smart numerical computations in chemical engineering courses using Mathcad is discussed to improve understanding of process fundamentals. The emphasis here is on obtaining solutions with increased confidence and higher precision and with minimum effort by understanding what type of problem is being presented. Students are encouraged to not introduce unnecessary complications when attempting to solve a problem and solve problems using optimal effort. Example problems considered for illustration deal with solution of linear algebraic equations (kinetic parameter estimation and properties of small and large reaction schemes and metabolic networks), nonlinear algebraic equations (vapor-liquid equilibria for ideal and non-ideal systems, equilibrium calculations for multiple reactions, isothermal and non-isothermal operations of well-mixed reactors with single and multiple reactions, and kinetic parameter estimation using nonlinear regression), and ordinary differential equations [ODEs] (isothermal, adiabatic, and non-adiabatic batch and steady state continuous flow tubular reactors), application of difference equations (population and system dynamics, stochastic processes), and symbolic algebra (rate expressions for catalytic reactions and characteristics of reaction networks). Illustrations involving ODEs pertain to initial value and boundary value problems. Analysis of multi-reaction networks is based on properties of atomic and stoichiometric matrices. A smart use of relations among atomic and molecular species and molecular species and reactions obviates the need for solving conservation equations for all species in the reaction network, which typically are systems of nonlinear algebraic equations and nonlinear ordinary and partial differential equations. Any independent reactions missing in the reaction network are identified in a systematic fashion. Reduced order reactor models for large reaction networks can be obtained by applying pseudo-steady state hypothesis for highly reactive species, such as free radicals and various forms of catalytically active sites, an approach also useful in flux balance analysis in networks of reactions in living cell cultures. Symbolic algebra is very useful in this effort. Mathcad is interactive, with user-friendly graphics capabilities. Students are encouraged to display results using graphics for improved learning and understanding. The capabilities of Mathcad are of significant benefit in accelerating the learning and strengthening the fundamental knowledge base.

494a Hands-on Hydro Proof Experiment Development

Gregory Ogden

This presentation will examine the details of a novel hydrostatic failure test developed for undergraduate chemical engineering unit operations courses. The experiment is was developed by both university and community college as part of a NASA MUREP project aimed at educating students through hands-on experiences and meet NASA education goals by strengthening the future workforce for NASA and the nation.

Initially, the hydrostatic test apparatus was created to support the development and testing of a 1U CubeSat green propulsion system including a polymer electrolyte membrane (PEM) electrolyzer, hydrogen and oxygen tanks and combustor. Each component of the propulsion process was custom made and required hydrostatic-pressure testing to establish operating safety margins. Furthermore, the Hydro Proof experiment contributes to the ever-increasing need for additional safety training in engineering curricula due to the significant numbers of industrial accidents each year. In 2020, someone died every 111 minutes from a work-related injury in the United States (Hess, 2021).

The experiment's background materials, pre-lab assignments and activities emphasizes the importance of lab safety education and prioritizes a hands-on approach. There are far too many examples of pressure-related accidents such as the accident in Florida where a pressure vessel ruptured and killed several employees (Johnson, 2008). There were car door-sized pieces of quarter inch-thick steel thrown several hundred yards from the locus of the blast (Figure 1).

The *Hydrostatic Failure Analysis Experiment* provides opportunities for hands-on experiences with safety testing techniques. As part of the experiment, students hydrostatically test commercial polyethylene (PET) bottles to failure. As part of pre-lab activities, students calculate hoop and axial stresses and explore the importance of safety factors in process operation. Following the experiment, they compare the theoretical predictions to experimental data.

The experimental equipment including pressurization pumps (pneumatic and battery powered), pneumatic controller, inlet water filter and hydraulic lines are presented in Figure 2. Images of the pre- and post-test water bottle are shown in Figure 3.

Topics include lab safety, equipment set up, pressure testing safety, and experimental procedures. Student data obtained during a beta-test run of the experiment from Fall semester 2023 will be reviewed. The pre-lab contains several questions regarding design safety, pressure testing and the advantages of using water as opposed to compressed air. This is followed by quick calculations to determine the equivalent force in the soda bottle failure when over pressurized with air to the explosive energy contained in Trinitrotoluene (TNT). Students were also asked to propose a 'dry lab-experiment' to visually compare the hazards/damage potential from an over pressurization event using compressed air vs. water in a one-and-a-half-foot cube. Initial student responses to this question will be presented.

References:

- Hess, Abigail, "13 U.S. workers die on the job per day, on average—and these are the most dangerous jobs", BLS estimates that 13 U.S. workers die on the job per day, on average (cnbc.com), accessed 3/11/2023
- Johnson, J. "Florida Explosion Investigation Yields Preliminary Results", C&EN, 2008. http://pubsapp.acs.org/cen/news/86/i02/8602news1.html, accessed 1/21/2023

Figure 1. Aftermath of runaway reaction explosion (Johnson, 2008)

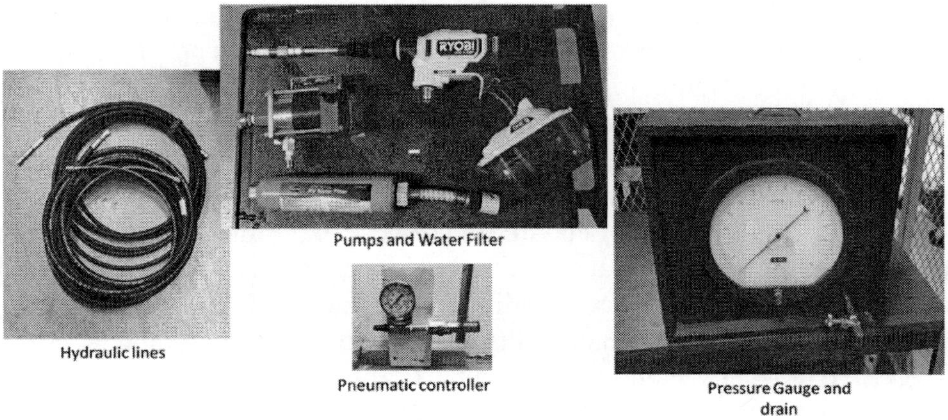

Figure 2. Hydrostatic Testing of PEM.

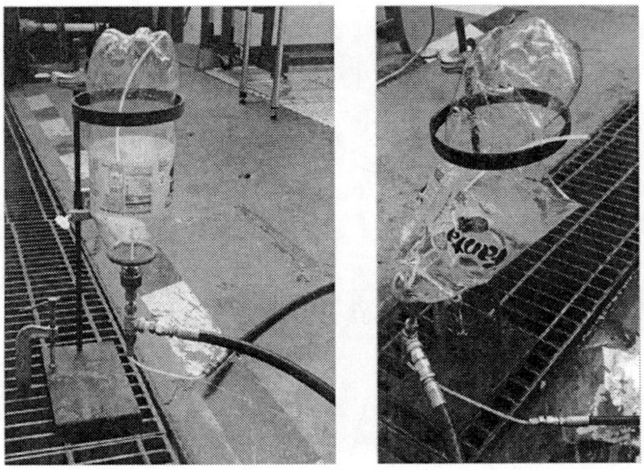

Figure 3. PET Water Bottle Test Images

494b Embedded Hands-on Classroom Activities Tailored to Personality Preferences Enhance Student Learning in Core Undergraduate Chemical Engineering Courses

Ifeoluwa Babalola, Jose C. Contreras-Naranjo, Vijay Ravisankar, Victor Ugaz

Laboratory courses in chemical engineering programs reinforce students' learning of fundamental theoretical concepts. Laboratory instruction, together with the use of computational simulation tools, enhance student mastery of fundamental concepts by connecting them with their practical application. However, lab courses are often challenging to deliver due to constraints of cost and scheduling of space and time. Additionally, the time-lapse between teaching theoretical concepts in core courses and performing the associated experiments in separate laboratory courses scheduled later in the curriculum greatly minimizes their impact. In this work, we explore optimal implementation of embedded experiential instruction to improve students' learning of core chemical engineering concepts. We designed in-class hands-on classroom activities for a junior-level chemical engineering fluid mechanics course that embedded in-class experiments, computational simulation, and a teaming guide tailored by personality preference. We adapted an experimental set-up, initially intended for laboratory use, into a hands-on kit for classroom use. The experiments were conducted with the hands-on kit safely in the classroom and required no extra space or time. We further explored the impact of team formation guided by personality preference based on the Myers-Briggs Type Indicator. After obtaining the experimental data, the students used commercial computational software (MATLAB) to analyze and interpret their results. The pre and post-test statistical analysis show a statistically significant improvement in students' learning of the fundamentals and the application of capillary flow. Our results indicate that incorporating low-cost and space-efficient hands-on experiments in classrooms, using personality preference as a teaming guide, can significantly improve students' learning of chemical engineering fundamentals.

494c A Product Design Workshop for Chemical Engineering and Chemistry K-12 Outreach

Stephanie Hill, Gregory T. Neumann, Adam J. Nolte, Daniel Anastasio

Many prospective college students have difficulty articulating the difference between chemistry and chemical engineering, and some even hold the misconception that they are the same discipline. To help students better understand the types of questions and approaches used by chemists and chemical engineers, we have developed a high-school outreach workshop demonstrating the intersection of science and engineering in the context of development of a personal care product. Our six-hour module is divided into three sessions, each approximately two hours. In the first session, students learn about how lotions interact with the skin and how different types of ingredients influence the final product's affective qualities. Later, students examine commercially available lotions and develop focus group questions. In the second session, students design a natural (non-petroleum) lotion and then implement their recipe to make a small amount of lotion in a chemistry lab. In the final session, students sample and rate each other's product using their focus group questions from session one. Following assessment of the subjective qualities of their lotions, students test viscosity, a quantitative factor, using a rotational viscometer. Finally, students are challenged to think about how to make their lotion in bulk quantities efficiently and given examples of how chemical processes can be scaled-up. The workshop is designed to mimic the process of designing a personal care product, emphasizing stakeholder input, formulation, testing (both affective and quantitative), and iteration. Additionally, the workshop is designed to be presented by both a chemist and a chemical engineer to highlight the role of each discipline in the process of bringing a product to market. We will share details of materials we selected for our lotions, possible alternatives, a summary of our experiences running the workshop, and ways that the workshop could be altered to meet the outreach needs of educators at other institutions.

494d Development of an Active Learning Methodology for Teaching Artificial Photosynthesis and Electrochemistry in Elementary Schools

Justin Bui, Brooke N. Soobrian, Alex J. King, Alexis T. Bell, Adam Weber, France A Houle

Artificial photosynthesis, which employs renewable energy sources to up-convert abundant feedstocks (*e.g.*, carbon dioxide and water) into value-added products, will be key to decarbonization of the energy and industrial sector in a sustainable future. Despite its importance, affordable and effective hands-on classroom activities that demonstrate artificial photosynthesis and teach key concepts in electrochemistry are lacking, particularly for primary school students. It has been well-documented that strong early-childhood education is key for encouraging positive career outcomes in the sciences, so educating primary school students on prospective sustainable energy technologies will be a critical step in the development of the next-generation energy workforce. All considered, there is a great need for a curriculum that introduces students to key sustainable energy technologies, such as artificial photosynthesis, earlier in their education.

In this talk, we present an artificial photosynthesis lesson plan based on an active-learning pedagogy that uses safe and highly accessible materials (baking soda, tap water, plastic jars, Ni coil, alligator clips, and a solar cell) to perform solar-powered water splitting. The lesson is developed in conjunction with established learning objectives in 5th-grade science education in the state of California, and compliments existing curricula on natural photosynthesis and engineering design. The lesson was presented in four 5th grade classrooms across the Bay Area, and the efficacy of the lesson plan in teaching basic concepts of artificial photosynthesis was evaluated with pre- and post-test data, which demonstrated a statistically significant improvement in overall student understanding. Importantly, the data showed that the lesson plan is effective at introducing students to electrochemistry concepts much earlier in their education. Lastly, we demonstrate that the active learning pedagogy employed in our lesson plan also helps to reduces the performance gap between over- and under-represented minorities, establishing that the developed lesson, and those like it, will be key in the development of a diverse energy workforce of the future.

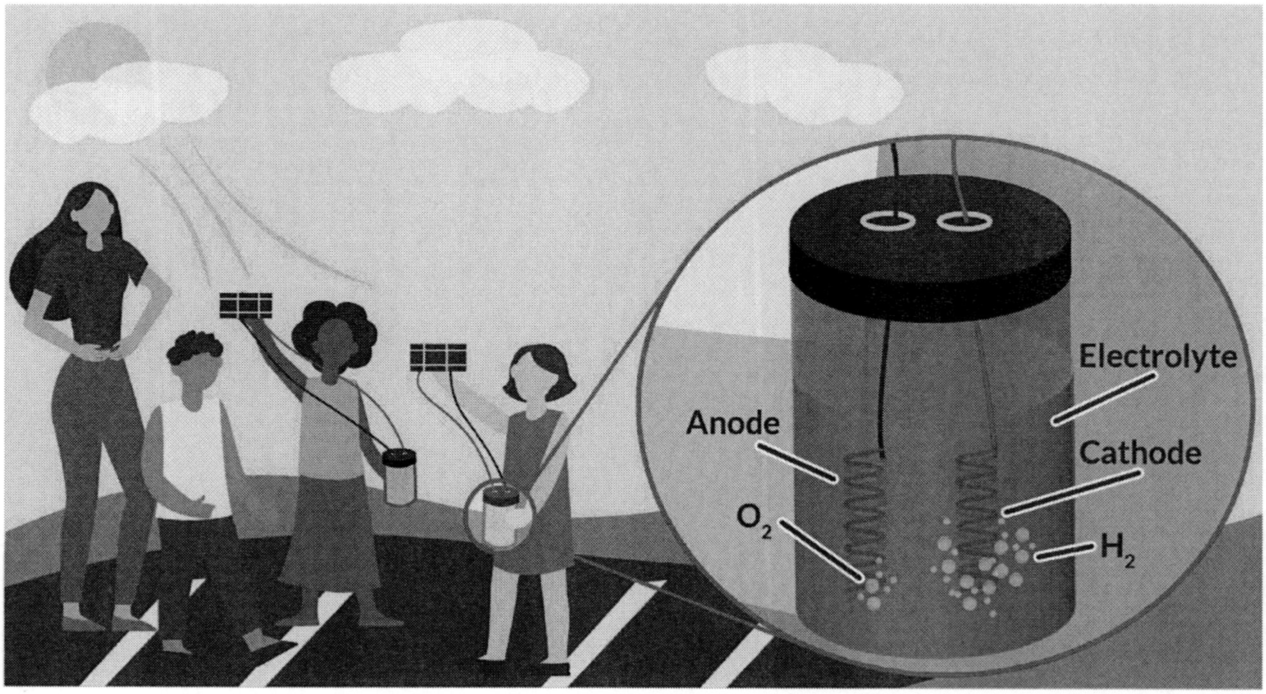

494e Using 3D Printing in an Introductory Chemical Engineering Course

Kristine Horvat

The 3D printing industry is expected to grow by up to 30% by 2026, and chemical engineers are well poised to benefit from this new technology. Therefore, exposing students to 3D product design and printing can be very beneficial. In order to introduce students to this field, chemical engineering students in a first year Chemical Engineering Technology course were tasked with designing, 3D printing, and testing a plastic tubing barb. First, students learned about process flow diagrams and AutoCAD design, and then they created their own 3D barb to connect two pieces of ¼" tubing using Inventor. Students were given a tour of the Makerspace, and they attended a short training session on how to use 3D printers. Students then printed their designs outside of class time, and they all brought their printed tubing barbs to class and tested them by connecting tubing and running water through the system. If a barb was found to leak or not fit the tubing, students were given a week to redesign and print a new barb, and then they tested out this new version. Upon completion of this activity, students wrote a reflection on the design, printing, and testing process. Overall, this was found to be an activity that students enjoyed while learning about 3D printing. This also proved to be a useful way to introduce learning from failure and the importance of redesigning and assessing a product.

494f Rig Building: A Fun Hands-on Lab Activity

Pavan Inguva, Andrew Macey, Umang V. Shah

Introduction

Student design competitions are a fun and effective experiential learning activity to achieve multiple learning outcomes. Students typically will have the opportunity to practice a range of technical and professional skills as they would need to work in a team to apply concepts and material from the classroom to complete a project that meets a minimum required scope by a defined deadline [1]. The value of design competitions for engineering education is well-known by educators and correspondingly, there are several competitions reported in the literature in various engineering disciplines across different scales (e.g., intra-department [2], inter-university [3], and even international [4]). However, the back-end of developing a student design competition can be complex and is often overlooked in the literature. Here, we describe the development and roll-out of a hands-on competition for first year chemical engineering undergraduates titled "Rig Building".

Context and Motivation

The MEng in Chemical Engineering program at Imperial College London is a four-year course leading to an integrated master's degree. In the first year, students take multiple introductory core chemical engineering classes such as transport, separations, and engineering mathematics. Students also complete the Foundation labs, an integrated practical module with a series of structured lab-based activities to help students develop essential skills such as safely manipulating equipment and analyzing experimental data [5]. Rig building was conceptualized as the final activity in the Foundation labs. As the last activity, it was the objective of the teaching staff to provide a fun and memorable experience to students while integrating the subject material covered in lectures and the Foundation labs. As design competitions are often very open-ended, it also helps set the tone for subsequent academic years as the work they will encounter will become increasingly unstructured.

Rig Building

Students are tasked to design and build a closed-system rig where the level and flowrate of water into a tank can be controlled independently. A list of available equipment and hardware for constructing the rigs is provided to students and much of the equipment is recycled year-to-year. Students have multiple constraints and requirements such as a set budget and design specifications that the rig has to achieve. Most constraints are posed as soft constraints where students are increasingly penalized for not meeting them. Students have access to a range of workshop services and expert help, however there is a cost associated with their use which counts towards the budget. The rig building competition lasts a total of one week with students having two days to construct their rig after submitting an initial design and risk assessment. Evaluations take place on the final day.

Assessment and Learning Outcomes

Prior to evaluation which is carried out by teaching staff, students have to compile and submit a series of documentation which includes a P&ID of their rig, a bill of materials and services used, an operating procedure, and an activity risk assessment. During the evaluation, students have to give a five-minute presentation to the assessors outlining the key features of the rig and demonstrate that is can meet the design specifications. Bonus points are awarded for creativity and novelty. Assessors are given a defined rubric to grade students' rig and deliverables. A small prize is awarded to the team with the best rig.

Historically, students have reported that rig building was the highlight of their Foundation labs experience with students reporting that the experience was fun and that it provided them a hands-on experience with tools and hardware. Many of the comments from course evaluations also indicate that students appreciated the open-ended nature of rig building in contrast to the typical Foundation labs project:

"The rig building was an excellent experience. Perhaps elements of that could be implemented in the actual labs"

"The rig-building project was excellent, as it was very fun and was a great experience to learn how to effectively use tools and equipment, and how to budget and plan well."

Conclusion

Rig building can be introduced as a fun and engaging small-scale design competition to younger undergraduate chemical engineering students. The project provides students an opportunity to practice multiple technical and professional skills expected from professional engineers and multiple learning outcomes can be explored in a comparatively short amount of time. Operationally, rig building can be implemented comparatively easily. It is also resource efficient as most of the equipment can be recycled year-to-year.

References

[1] C. Kaiser and W. Troxell, "Student Design Competitions in Undergraduate Engineering Education," in *Proceedings Frontiers in Education 35th Annual Conference*, 2005, pp. S3J-13-S3J-15, doi: 10.1109/FIE.2005.1612297.

[2] S. Mandavgane, "Fun with fluid: An innovative assignment in fluid mechanics," *Educ. Chem. Eng.*, vol. 30, pp. 40–48, Jan. 2020, doi: 10.1016/j.ece.2019.11.001.

[3] AIChE, "Chem-E-Car Competition," 2023. https://www.aiche.org/students/chem-e-car-competitionr (accessed Mar. 17, 2023).

[4] M. Gadola and D. Chindamo, "Experiential learning in engineering education: The role of student design competitions and a case study," *Int. J. Mech. Eng. Educ.*, vol. 47, no. 1, pp. 3–22, Jan. 2019, doi: 10.1177/0306419017749580.

[5] W. Chen, U. V. Shah, and C. Brechtelsbauer, "A framework for hands-on learning in chemical engineering education—Training students with the end goal in mind," *Educ. Chem. Eng.*, vol. 28, pp. 25–29, Jul. 2019, doi: 10.1016/j.ece.2019.03.002.

494g Increasing Student Engagement in Chemistry Via Incorporation of a Catalyst Design in-Class Activity

Alyssa Hensley, Dawn Shum, Patricia Muisener, Rahul Khade, Jeffrey Raab, Anju Sharma

While students in introductory chemistry classes at both the high school and undergraduate levels are often highly engaged when discussing topics such as climate change, there is an obvious disconnect between such grand challenges and how the fundamental chemistry topics and research apply to such challenges. This disconnect between the topics discussed in class and their application to worldwide challenges discourages students from participating in class. Thus, it is of critical importance to engage students in open-ended and design-based problem-solving with chemistry.

Here, we present two case studies, one at the high school level and one at the undergraduate level, of an open-ended and design-based problem-solving approach implemented in chemistry classrooms using the topic of heterogeneous catalyst design for hydrogen fuel cells. Hydrogen fuel cells are an emerging technology that produce electricity from hydrogen oxidation to water. Such a process is environmentally beneficial due to the lack of the carbon-based byproducts that are currently contributing to ever-harshening climate change consequences. However, a major drawback of hydrogen fuel cells is the catalyst material required for such a process, i.e. Pt. This provides a concrete research question to engage high school students with: how can we replace the expensive and unsustainable Pt catalyst in hydrogen fuel cells?

To address this research question, we have developed a catalyst design graphical user interface, named Computational Catalysis Design Interface (CCDI), and two lesson plans (high school and undergraduate) that enable students to explore and design new materials for an emerging carbon-free route to produce energy. This presentation will show you how to engage students in their exploration of the core concepts related to hydrogen fuel cells, heterogeneous catalysts, the influence of catalyst properties on performance, and how key chemistry concepts (i.e. thermodynamics) can be applied. Through this activity, students gain hands-on experience in applying the scientific method to today's critical challenges as well as begin to see themselves as a scientist.

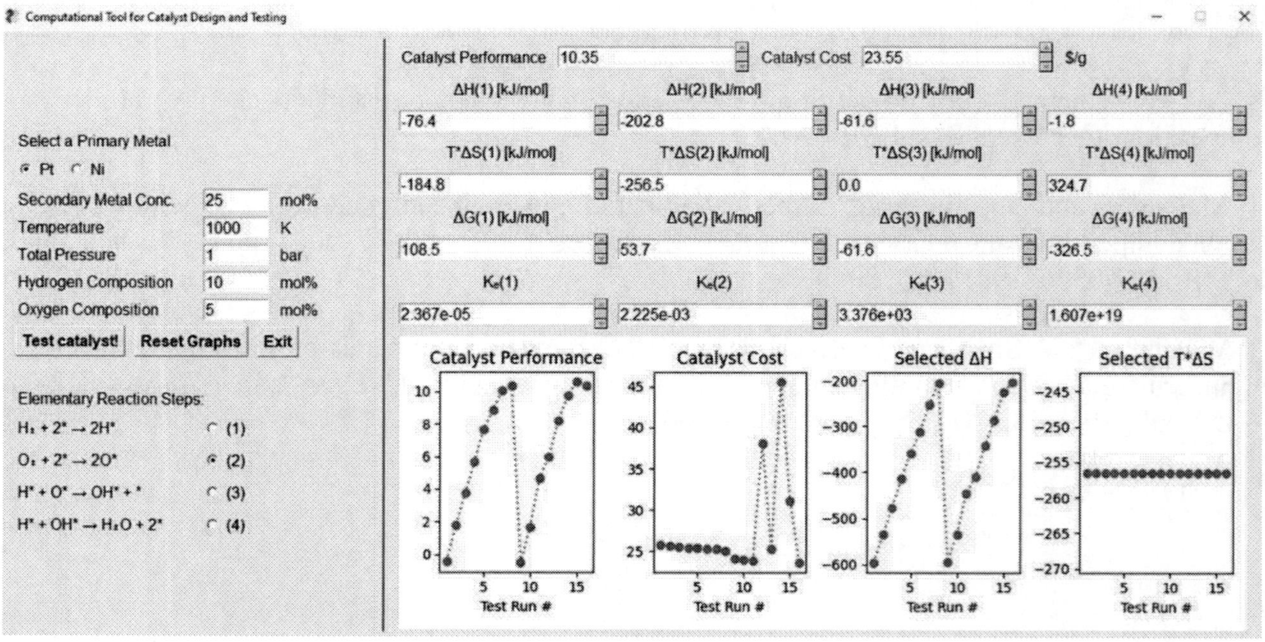

494h *Teaching Process Safety and Control through a Real-Time Approach to Industrial Settings*

Stefanos (Steve) Nitodas

Despite the significant benefits that society has gained from process control technology and the continued vitality of control science as a research field, the academic research community and the practitioners of process control in the industry are insufficiently connected to each other [1]. Developing an engaging process safety course at the university level should not rely only on material obtained from textbooks, but also on the hands-on experience that the instructors have obtained from the industry and real production plants. A good process control is synonymous with good chemical engineering and this relation should be demonstrated through actual case studies that the instructors have experienced in industrial settings.

For decades, the subject of control theory has been taught using transfer functions, frequency-domain analysis and Laplace transform mathematics. This approach is proper for linear systems. However, in the control of chemical processes, which are often characterized by nonlinearity and large doses of dead time, classical control techniques have some limitations. As a consequence, the chemical engineering students do not need the detailed theory and mathematics that usually surround the teaching of process control, such as frequency domain analysis. Instead, they need a base level of understanding of differential equations, as well as comprehension of process dynamics, dynamic modeling of the main unit operations in the time domain, basic control algorithms and cascade structures that are often encountered in the industry. With these basic tools and the perception of how to apply them, the chemical engineers can solve most of the control problems themselves.

The basis of this study is to provide a practical, hands-on introduction to the topics of process control by using only time-based representations of processes and the associated instrumentation and control. In the past four years, a process control and safety course has been developed that is based on fifteen (15) years of practical experience of the instructor from the industry, considering the aforementioned needs of the chemical engineering students to understand process control in real-time environments. The course considers the investigation of processes initially in the Laplace domain in order for the students to understand and master classical process control techniques. Subsequently, the study of the same processes in the time domain is carried out, thus connecting the students with the respective equations and techniques employed in the industry, focusing on introducing the very important and often overlooked topic of instrumentation, and control-loop design and tuning. Advanced control configurations are also considered, including split control and override control.

The right combination of hardware and software is used in the course to implement a 'hands-on' approach to process control system design. In this way, engineers and students alike will be able to experiment on virtual plants that capture important nonideal cases of the real world, and readily test even the most outlandish of control structures without resorting to non-intuitive mathematics or to placing real plants at risk [2]. This course has also considered integrating process safety and associated cost that are important for the global understanding of process control by the future chemical engineers. Finally, a combination of theory and applied methodology from real examples from the industry is used to provide a practical treatment of process control.

References

[1]. Tariq Samad, Margret Bauer, Scott Bortoff, Stefano Di Cairano, Lorenzo Fagiano, Peter Fogh

Odgaard, R. Russell Rhinehart, Ricardo Sánchez-Peña, Atanas Serbezov, Finn Ankersen, Philippe Goupil, Benyamin Grosman, Marcel Heertjes, Iven Mareels, and Raye Sosseh, "Industry engagement with control research: Perspective and messages", *Annual Reviews in Control*, Vol. 49, 2020, pp. 1-14, https://doi.org/10.1016/j.arcontrol.2020.03.002.

[2]. William Y, Svrcek, Donald P. Mahoney, D.P., and Brent R. Young, *"A Real-Time Approach to Process Control"*, , 3rd ed., John Wiley & Sons, Ltd., 2014. ISBN- 13: 9781119993872.

494i Design and Development of an It / OT Emerged Experiment for Industrial Cybersecurity

Sharath Daida, Helen Lou, Tianxing Cai

The increasingly complex and networked characteristics combined with industrial control system (ICS) make processes vulnerable to cyber-attacks that can lead to major disruptions, or even incidents and accidents. In this study, a miniature physical system was designed and developed to mimic a typical liquid handling process in the energy infrastructure. This miniature system includes storage devices, pumps, and control valves. The process has been built at the bench-scale for easy demonstration of research activities and the microcontroller has been built by using Arduino kit. Different industrial cybersecurity experiments can be conducted using this platform.

554a The Prevalence of Impostor Phenomenon in Chemical Engineers

Sarah Wilson, Luke Landherr, Tracy Carter, Marion Carter

Impostor phenomenon is a psychological experience where individuals feel that they are not as successful as others believe them to be. This can leave them feeling like a fraud, feeling like they have only obtained their current position out of luck or fearing that one day others will find out that they don't belong. While the phenomenon was first characterized in high achieving women, it has since been shown to impact individuals across all identities. Within chemical engineering, Richard Felder first commented on the impostor phenomenon in a 1980 article in Chemical Engineering education. Recently, an article in Science talked about the potential impact that impostor phenomenon could have on pathways through STEM for students who are historically underserved in engineering. As a result, there is a need to better understand the prevalence of impostor phenomenon within the chemical engineering community. Through this study, survey data was collected from over 500 chemical engineering students and professionals (in both industry and academia). Results show that 74% of undergraduate and graduate students in chemical engineering experience frequent or intense impostor characteristics. This number is decreased to 55% for chemical engineers in faculty or industry roles. Additionally, impostor characteristics were found to decrease with age, with only 32% of individuals over the age of 55 experiencing frequent or intense impostor characteristics. This study highlights the significant number of chemical engineering students and professionals who are experiencing feelings of impostor phenomenon. It further motivates the need for research on reducing the impact of impostor phenomenon on individuals, especially those from marginalized backgrounds, as they navigate their education and careers in chemical engineering.

554b From Advocacy to Action: Mac's Efforts to Serve Underrepresented Groups in Chemical Engineering through Latinx in ChE Affinity Group

Sindia M. Rivera-Jimenez, Gabriel Rodriguez, Fabiola Capa Villarreal, Viviana Monje-Galvan, Isabela Ramirez Velez, Eduardo A. Chacin Ruiz, Roque Gochez, Alejandro Forigua Coronado

This presentation describes the formation and structure of the LatinX in ChE affinity group under the Minority Affairs Committee (MAC) from the Societal Impact Operating Council (SIOC) at AIChE. The presentation will discuss how MAC, historically focusing on improving the representation of Black engineers for over 30 years, has advocated for expanding its reach to other underrepresented groups in chemical engineering. With the implementation of the AIChE IDEAL Path, MAC has initiated changes in its bylaws to include a new community component known as affinity groups to increase inclusivity for all interested members. The reorganization of MAC as a community aims to serve better racial and ethnic groups that are underrepresented in the chemical engineering profession.

The LatinX in ChE Affinity group is one such group created as a grassroots effort of volunteers within MAC and SIOC to connect members of Latin American descent. Since 2022, the group has organized various activities to support the success of chemical engineering students and professionals at all levels. These activities include career development workshops, networking events, and discussions on topics related to diversity, equity, and inclusion in chemical engineering. A unique aspect of the scalability and sustainability of LatinX in ChE is that the leadership board comprises students, faculty, and industry professionals, providing a space for dialogue and collaboration. The presentation will discuss best practices for creating affinity groups under MAC that promote culturally relevant educational activities and professional development opportunities, aimed at empowering other underrepresented groups.

554c An Inclusive Teaching Practice Intervention in Two Graduate Chemical Engineering Classes

Prajnaparamita Dhar, Meagan Patterson

There is a critical need for more, and more diverse, engineers in the U.S. (Anderson et al., 2018; Brooks, 2020). Lack of representation in engineering fields increases as level of education increases (i.e., engineers with doctoral degrees are a less racially and ethnically diverse group than those with bachelor's degrees; Anderson et al., 2018). Creating a more diverse engineering workforce begins with increasing the diversity of engineering students. Many attempts to increase the diversity of the engineering student population focus on recruitment, but retention of students is also critical (Woolston, 2021). Many retention programs are grounded in a student affairs perspective and emphasize support and engagement outside of the classroom (e.g., student affinity groups). Particularly at the graduate level, however, positive classroom experiences are critical to student success. The specific goal of this work was to promote inclusive teaching practices in a chemical engineering graduate program, with a particular focus on two core chemical engineering graduate courses (a required technical course for first year graduate students and a required seminar course for all graduate students in the program). Specifically, changes were made to the syllabus to make it more anti-racist, by following established best practices in syllabus design. More active learning activities and group discussions were designed and implemented and will be discussed in more detail.

Data were collected to examine the impact of this microintervention on students' self-reported perceptions of department climate, classroom engagement, and sense of belonging. In order to examine the impact of the microintervention on students, we compared pre-intervention data (from a survey of the full department in February 2022) to post-intervention data (from a survey of students in one of the core intervention courses in May 2022). Results indicated that compared to the full department, students in the intervention group reported similar levels of classroom engagement and sense of belonging, but had more positive perceptions of climate within the department. These finding suggest that even relatively minor changes to teaching in core graduate courses can impact students' feelings of inclusion within their graduate programs, and that more intensive interventions might have even larger impacts on student attitudes and retention.

References

Anderson, E. L., Williams, K. L., Ponjuan, L., & Frierson, H. (2018). The 2018 Status Report on Engineering Education: A Snapshot of Diversity in Degrees Conferred in Engineering. Association of Public & Land-grant Universities. https://www.aplu.org/library/the-2018-status-report-on-engineering-education-a-snapshot-of-diversity-in-degrees-conferred-in-engineering/File

Brooks, J. (2020). Why Should I Care About Diversity in Engineering? National Society of Professional Engineers. https://www.nspe.org/resources/pe-magazine/july-2020/why-should-i-care-about-diversity-engineering

Woolston, C. (2021). Minority representation in US science workforce sees few gains. Nature, 592, 805-806. https://doi.org/10.1038/d41586-021-01089-6

554d Preliminary Findings from Research on Inclusive Mentoring Strategies for Neurodivergent Undergraduate Researchers in STEM

Mariah Arral, Elise Baribault, Sarah Young, Julianna Gesun, Jeffrey Halpern

Undergraduate research opportunities improve retention, increase confidence, and support better overall academic performance for underrepresented students in STEM. Neurodiversity is a term used to encompass all individuals, inclusive of both neurodivergent and neurotypical. The neurodivergent community is often inappropriately perceived to have disadvantages associated with STEM-based research and faces exacerbated challenges when pursuing undergraduate research with STEM faculty. Despite the increase in STEM students who report disabilities, few resources are available to train mentors to work with this population. Our team seeks to investigate different mentoring strategies that support neurodivergent STEM undergraduate researchers to thrive. The aim is that this project will provide educators with information that can improve how we attract, support, retain, and graduate a neurodiverse population of engineering students. Additionally, increasing the success of neurodivergent STEM undergraduate students through mentorship not only broadens participation in STEM but also provides more role models for current and future students. To assess our aims, we created a survey for the undergraduate research community. After the publication of a work-in-progress paper in the summer of 2022, we updated our survey questions from feedback at the American Society for Engineering Education annual meeting, this was followed by the conduction of cognitive interviews on our assessments. From these cognitive interviews, our quantitative questions were again updated. Then, the survey was sent out to undergraduate students across STEM disciplines. Both neurotypical and neurodivergent students were surveyed. Survey questions explored students' experiences with mentorship, research, thriving, and background information. This presentation details the quantitative results from the survey questions with engineering students who participated in undergraduate research. Future work will detail interviews with students and faculty on their experiences with mentorship and the STEM research environment.

554e Exploring the Use of Natural Language Processing in Developing Problem-Based Learning Scenarios for Social Responsibility in the Curriculum.

Sindia M. Rivera-Jimenez, Kassandra Fernandez

Social responsibility in engineering refers to the ethical obligation of engineers to consider the broader impacts of their work on society and the environment. Incorporating social responsibility themes into the engineering curriculum can encourage students to think critically about the societal implications of their work, including environmental sustainability and social justice. However, creating new problem-based learning (PBL) scenarios that incorporate social responsibility themes can be challenging for many faculty due to limited time and resources, lack of training and support, limited access to diverse perspectives, and balancing social responsibility with technical content.

This work explores the use of Natural Language Processing (NLP) as an instructional tool to develop PBL scenarios using social responsibility themes in an introductory chemical engineering course. NLP-based topic analysis tools can analyze large amounts of text from various sources to identify key topics and subtopics related to social responsibility, helping faculty to identify important themes to incorporate into PBL scenarios. Additionally, NLP-based text clustering tools can group similar texts related to social responsibility, allowing faculty to identify patterns and develop PBL scenarios that are relevant to current social issues. These NLP-based tools can aid in the development of PBL scenarios that encourage critical thinking and engagement among students in social responsibility themes, preparing them to be more responsible engineers. The effectiveness of the PBL statements in promoting critical thinking skills among chemical engineering students will be evaluated and compared to traditional PBL statements. Pedagogical strategies, challenges and ethical considerations for using AI tools and techniques in engineering education will also be presented.

554f Incorporating Environmental Justice into the Mass and Energy Balances Curriculum

Jessica Winter

With increasing societal interest in chemical safety and the potentially harmful effects of chemical manufacturing and transportation, there is a desire to include environmental considerations within the chemical engineering curriculum. This case study evaluated the intersection of environmental pollution as represented by regional particulate matter 2.5 (PM2.5) levels, population disparities in pollution exposure based on localization to chemical manufacturing facilities, and Covid-19 mortality rates. The curricular content used 3 peer-reviewed journal articles[1-3] that examined population disparities in pollution exposure and their correlation with Covid mortality rates in the US and specifically in Louisiana's "Cancer Alley" region, home to a high concentration of chemical manufacturing facilities. Student engineering learning outcomes included practice using unit conversions, increased understanding of environmental regulatory guidelines and definitions (e.g., PM2.5), and practice interpreting statistical correlations in data sets. In addition, student outcomes included public health and safety and ethical practice of engineering content. Over the duration of course engagement, students increased their understanding of air pollution and its sources and its correlation to health effects (from 86% mastery to 100% mastery over 3 assignments). Student ability to correlate these with population disparities was highly variable throughout the assignments (from initial 21% to 100%, mode 50%), suggesting an area for improvement. Nonetheless, this case study provides a starting point for integrating environmental justice considerations into the chemical engineering curriculum in support of the ethical practice of the profession.

1. C. W. Tessum, J. S. Apte, A. L. Goodkind, N. Z. Muller, K. A. Mullins, D. A. Paolella, S. Polasky, N. P. Springer, S. K. Thakrar, J. D. Marshall and J. D. Hill, *Proceedings of the National Academy of Sciences*, 2019, **116**, 6001-6006.
2. K. A. Terrell and W. James, *Environmental Justice*, 2020, DOI: 10.1089/env.2020.0021, pre-print.
3. X. Wu, R. C. Nethery, M. B. Sabath, D. Braun and F. Dominici, *Science Advances*, 2020, **6**, eabd4049.

554g From Research to Advocacy: A Workshop to Develop the Next Generation of Socially Responsible Scientists and Engineers

Kassandra Fernandez, Sindia M. Rivera-Jimenez

Professional organizations like the Accreditation Board for Engineering and Technology and the National Academy of Engineering have emphasized the importance of social responsibility for modern engineers. Scientific communication and policy advocacy are promising teaching approaches to foster the development of critical thinking skills in undergraduate researchers related to social responsibility. Scientific communication, or SciComm, is the practice of turning scientific knowledge into accessible and engaging content geared towards the general public. By making empirical data and reasoning more understandable, SciComm can help people make better decisions and inform public policymaking. Policy advocacy involves actively supporting and engaging with specific policies. Given their specialized knowledge and expertise, engineers are well-suited to engage with the public and promote science-based policies.

To instill the importance of critical thinking and social responsibility, engineering students should learn about SciComm and policy advocacy starting in their undergraduate years. In this presentation, we will describe the results of a two-day workshop that took place during the Summer of 2023. This workshop aimed to create awareness of the role of scientists and engineers in policy decision-making, and to develop effective communication strategies to engage with the public on policy issues. On Day 1, the workshop covered the importance of scientists and engineers in policy decision-making, effective communication strategies, and group activities to craft a clear and compelling message for policy communication. On Day 2, participants engaged in a group practice session where they learned to communicate complex scientific information in accessible terms. This session included feedback on the participants' communication skills. The workshop concluded by providing sources and materials on effective communication strategies and policy decision-making for further learning and development. Preliminary data was collected before and after the workshop took place using the VSRoSE validated survey instrument and open-ended questions. This work is part of a major study exploring the importance of social responsibility in undergraduate engineering research experiences.

577a Integrating Sustainability into Chemical Engineering Education: Experiences from Imperial College London

Umang V. Shah, Pavan Inguva, Jerry Heng

Introduction

Chemical engineers are uniquely positioned to help address many of the sustainability challenges faced today. Many core chemical engineering topics such as separation processes and process control and optimization are highly relevant to ensuring the technical success of a sustainability related endeavor. Chemical engineers also are expected to understand process economics and financials which are crucial for the commercial viability of any project. This has resulted in many chemical engineers being at the forefront of many sustainability projects in many fields. Considering the importance of sustainability and that the job market for green jobs is growing rapidly, it is imperative that educators provide suitable opportunities to help graduates develop the right skillsets and attitudes to succeed in a green career and incorporate sustainability into their professional practice regardless of field. We outline how sustainability is conceptualized and integrated into the chemical engineering curriculum at Imperial College London as part of a periodic curriculum review process conducted at the college. We hope to share best practices and foster a conversation about how sustainability can be incorporated into the chemical engineer's educational journey.

Conceptualizing Sustainability in Chemical Engineering

Sustainability in the context of chemical engineering education has multiple dimensions to be considered. From a content perspective, chemical engineers have a role to play in many areas such as such as carbon capture, utilization, and storage (CCUS), renewable energy and energy storage, recycling and upcycling, water management, and policy planning [1]. These areas of research and practice draw upon both core topics and emerging research themes in chemical engineering. Within a given field, the type of work which can range from basic research to process design and techno-economic analysis needs to be accounted for. Educators also have to consider the broader sustainability paradigm and incorporate learning activities to help students develop the right attitudes and soft skills such as holistic thinking (both vertically e.g., from a global to individual perspective [2] and laterally e.g., across disciplines and domains [3]) and experience working in diverse multidisciplinary teams.

Leveraging Our Strengths

Sustainability has historically been and remains an area of key focus for the department. Many of the faculty and staff are world leaders in research areas aligned with sustainability such as energy storage and management, CCUS, green chemistry, process systems engineering, and techno-economic analysis. Some of the faculty are also affiliated with other sustainability efforts across the college such as the Grantham Institute and the Energy Futures Lab. Various collaborations spearheaded by the department such as the Sargent Centre for Process Systems Engineering and Institute of Molecular Science and Engineering The deep expertise within the faculty and staff in the department has provided a robust foundation upon which many modules, projects, and learning activities are built on.

Considerations for Incorporating Sustainability

Exploring the integration of sustainability into the chemical engineering program requires considering the requirements and limitations of various stakeholders. The type and level of learning activity developed for students must be commensurate with their educational experience and academic / professional interests. While some content will be deemed core and needs to be introduced to students early on, other content maybe too advanced or specialized and better suited as material for electives or advanced projects. The scope, structure, and scaffolding provided to students for a given learning activity will need to be planned for accordingly. Educators also need to be mindful that contact time with students is limited. This point is also related to the mechanism by which sustainability is integrated. This typically takes place in a few well-established modes: 1. Embedding sustainability concepts into a regular course, 2. Developing a new course, 3. Providing a sustainability specialization pathway within the degree, and 4. Developing an active learning activity [4].

Ensuring alignment with accreditation requirements presents another dimension to consider. While various accreditation boards have incorporated sustainability in one form or another into the expected student outcomes, multiple issues can arise including balancing an increased focus on sustainability with other technical requirements [4]. On a related note, alignment between the curriculum and industry is important, but can be challenging to achieve.

Incorporating Sustainability at ICL

We breakdown how sustainability is integrated into the chemical engineering curriculum from the 1st to 4th academic years. In the 1st year when students are typically covering chemical engineering fundamentals, sustainability is often introduced into core classes as a context for application of the material covered. An example is the coursework for an introductory physical chemistry course where students explore the properties of CO_2 and how it is a greenhouse gas. As students progress academically, a variety of learning activities with a greater focus on sustainability can be introduced. This includes experiential learning projects such as comprehensive lab projects where students explore sustainability concepts like operating a water purification process or process intensification. A highlight in the undergraduate curriculum is the Pilot Plant project where 2nd year students operate and optimize a state-of-the-art carbon capture pilot plant built in collaboration with ABB [5]. A range of courses, either incorporating sustainability concepts into the course or having sustainability as the main focus, are available for students. In the 3rd and 4th years, students can take multiple electives on sustainability which include courses on carbon capture and green fossil fuels, sustainable energy technologies, and the techno-economics of climate change solutions. Students are also required to take a course on environmental engineering which explores a range of topics such as geochemical cycles, pollutant transport and meteorology, and emissions control. Coursework components of other courses can also become increasingly sophisticated, incorporating techniques from current areas of research such as employing molecular design techniques to identify environmentally-friendly refrigerants in an optimization course.

Conclusions

IIntegrating sustainability into the chemical engineering curriculum is vital to prepare students to engage with intricate environmental problems and address global sustainability challenges. The involvement of stakeholders not only ensures the implementation of sustainable and realistic engineering practices but also affirms the authentically and holistically incorporation of these practices across the department. Several dimensions, such as the learner's academic maturity, module context, and the desired learning outcomes, must be considered to design the course elements and adequately outline the intended learning outcomes. By undertaking these strategies, the existing resources and expertise within the department and across the college can be leveraged to help create an impactful educational experience for students that incorporates state-of-the-art material and techniques.

References

[1] H. Arastoopour, "The critical contribution of chemical engineering to a pathway to sustainability," *Chem. Eng. Sci.*, vol. 203, pp. 247–258, Aug. 2019, doi: 10.1016/j.ces.2019.03.069.

[2] R. J. Batterham, "Sustainability—The next chapter," *Chem. Eng. Sci.*, vol. 61, no. 13, pp. 4188–4193, Jul. 2006, doi: 10.1016/j.ces.2005.10.016.

[3] A. Argoti, A. Orjuela, and P. C. Narváez, "Challenges and opportunities in assessing sustainability during chemical process design," *Curr. Opin. Chem. Eng.*, vol. 26, pp. 96–103, Dec. 2019, doi: 10.1016/j.coche.2019.09.003.

[4] M. Thürer, I. Tomašević, M. Stevenson, T. Qu, and D. Huisingh, "A systematic review of the literature on integrating sustainability into engineering curricula," *J. Clean. Prod.*, vol. 181, pp. 608–617, Apr. 2018, doi: 10.1016/j.jclepro.2017.12.130.

[5] ABB, "Imperial College carbon capture pilot plant: Preparing today's students for tomorrow's world," Cambridgeshire, 2012.

577b Teaching Sustainability to Undergraduate Students: Predicting Life Cycle Inventory Data for Environmental Impact Assessment

Austin Lehr, Emmanuel Aboagye, Elizabeth Dellorco, Ethan Shumaker, C. Stewart Slater, Mariano J. Savelski, Robert Hesketh, Kirti Yenkie

Advances in the chemical industry have led to the discovery of new compounds that do not have documented properties. Due to their novelty, access to their Life Cycle Inventories (LCIs), environmental impacts, and other sustainability metrics can be challenging as there is little available information in the literature. Furthermore, there are existing chemicals without such data available. Predicting these metrics are essential in assessing the sustainability of novel and existing chemicals. Experimentation or computationally intensive large-scale computations such as Density Functional Theory (DFT) and Molecular Dynamics (MD) are time and energy intensive which prevents facilities from performing Life Cycle Assessments. Since LCI data are required to perform any environmental impact assessment, it is imperative to find ways for its estimation to enable the development of greener processes.

In this work, we use Machine Learning (ML) methods to predict LCIs readily from molecular descriptors and thermodynamic properties. We first construct a database containing the properties of existing chemicals, such as density, molecular weight, boiling point, functional groups, number of atoms, toxicity, and many more. Overall, 23 and 200 thermodynamic and molecular properties were gathered respectively. We then use the Ecoinvent database from SimaPro® and the National Renewable Energy Laboratory's United States Life Cycle Inventory database to gather LCI data for the existing chemicals. The gathered data is then used to build a supervised learning model using the gradient boosted trees algorithm in python. This model will be used to predict several different impact categories, human health, ecosystem quality, climate change, and resource requirement. Human health impact is measured in Disability Adjusted Life Years (DALY), ecosystem quality impact is measured in the Potentially Disappeared Fraction of a species in a given area per year (PDF*m^2*year), climate change is measured in kilograms of carbon dioxide equivalent (kg CO_2 eq), and resource requirement is measured in megajoules (MJ) .

In this work we explore the teaching of sustainability assessment and design through the Engineering Clinic course offered by the Henry M. Rowan College of Engineering at Rowan University. The Engineering Clinic is a 2-credit course offered every semester with 3-hour meetings scheduled twice a week. Engineering Clinic teams are comprised of undergraduate students, graduate students, and faculty advisors. The faculty advisors' network with industries and research institutions and obtain grants for this work from federal and regional organizations as well as industries. In addition, the faculty advisors define the overall goals, tasks, and learning outcomes of the project. The graduate students are responsible for the dissemination of the tasks defined by the faculty and mentorship of the undergraduate students. The undergraduate students meet twice weekly with the graduate students and faculty advisors to share their work progress and receive feedback and directions for the next steps. The deliverables for the undergraduate team include a final comprehensive report and technical presentation towards the end of the academic semester. Using a well-established rubric for final grading, the students are assessed based on their overall performance in the project in regards to initiative, clear problem definition, identification of solution methods and application, technical presentations and final report. During the presentation, students are evaluated on various categories such as project introduction effectiveness, organization and clarity of presentation slides, visual aids for effective communication, and overall handling of discussions. Furthermore, many students also participate in regional and national conferences to present their work, thereby enhancing their professional and networking skills.

This undergraduate-graduate-faculty synergy allows for sufficient technical input from the faculty advisors, an opportunity for the graduate students to interact with stakeholders, and the undergraduate students a chance to apply their classroom knowledge to real-world engineering problems.

577c Preparing Future Engineers to Tackle Global Challenges: Fostering Global Awareness and Social Responsibility through Capstone Design Course

Courtney Pfluger, Betul Bilgin

As the world becomes increasingly interconnected and global issues continue to arise, it is crucial to have engineering students understand the complexities of solving real world global challenges as they will be the ones innovating solutions for the future. There is a need to understand how to better provide students with learning experiences that provide them opportunities to analyze and evaluate engineering concepts to a design solution as well as the environmental, social, cultural, and economic ramifications of their proposed solutions. Translating these types of learning outcomes to the classroom is important as educators to give the students authentic experiences and projects to facilitate this cross-cultural learning and informing students of the global context to large societal issues and fostering social responsibility and awareness. This is especially true in the capstone design course, which is the cumulation of the undergraduates' engineering experience. This team design project relies on student's connecting their experiences, curriculum knowledge, and skills to design a safe, efficient, and feasible process. Specifically, in chemical engineering their capstone projects focus on process and plant design, which can have environmental, social, and economic issues that can affect the project design and decision making. There is a need to understand and foster student decision making surrounding social responsibility and global awareness in their project designs using established theoretical models and frameworks. The OECD Program for International Student Assessment (PISA) global competence framework and questionnaire was used as a framework to assess how students perceived these global and societal issues in their projects. The PISA framework has four pillars: 1. the capacity to examine issues and situations of local, global, and cultural significance; 2. the capacity to understand and appreciate different perspectives and worldviews; 3. the ability to engage in open, appropriate, and effective interactions across cultures; and 4. the capacity to take action toward sustainable development and collective well-being.[1] The students were given the voluntary questionnaire before and after the capstone design course which included the PISA global competence questions as well as information on demographics, languages spoken, and personal cultural influences.

This talk will discuss how the four pillars of global competency were implemented in the senior capstone design course at two universities. It will address best practice topics such as lectures that help students understand how to implement sustainable and safety practices to ensure responsible process design, project topics and selection, location selection, and social, cultural and economic decision making for location. The impacts of these best practices on students' awareness of global and societal issues will be discussed with the lens of PISA framework.

1. OECD (2019), 'PISA 2018 Global Competence Framework', in PISA 2018 Assessment and Analytical Framework, OECD Publishing, Paris, https://doi.org/10.1787/043fc3b0-en.

577d Teaching Systems Thinking for Sustainability - Lessons Learned from the Project-Based Course 'the Global Energy Future"

Marcello Di Martino, Bassel Daher, Efstratios Pistikopoulos

The project-based course "The Global Energy Future" brings together interdisciplinary teams of students to address future energy challenges through a food-energy-water nexus (FEWN) lens. The course explores the interlinkages between the various resource supply sectors in the Houston region and their impact on sustainable development [1, 2]. A key focus of the course is the development of scenario analysis and decision support tools to guide better decision making for the energy future of Houston. The course also emphasizes an interdisciplinary approach, including data analysis, scenario building, and teamwork, to address the complex issues of energy futures. The key learning outcomes for the students are to (i) apply critical thinking and problem-solving skills to analyze complex systems and develop recommendations for addressing issues related to global energy futures, (ii) develop skills in data collection and analysis, including the use of appropriate software and tools, (iii) understand the importance of using data to inform decision-making and recommendations related to the FEWN, (iv) develop skills in integrating different perspectives and knowledge from different disciplines to inform the analysis and recommendations.

To this end, students are divided into different groups based on their background. According to the works of Daher et al. [3, 4] each group has to define the critical question of their project, the relevant scales and the stakeholders. In the next step, each group has to identify the key system of systems to be included and their interconnections, the assessment criteria, what type of data is needed and how to involve decision-makers in the process. Ultimately, each group has to derive a quantitative tool answering their own individually derived critical question incorporating holistic systems thinking for a sustainable food, energy, water resource supply of Houston. The project topics of last semester's class include (i) How to increase Houston's water supply by 25% by 2050; (ii) Can Houston be powered 50% by renewables in 2050; (iii) Techno-economic feasibility analysis of seawater desalination paired with renewable energy sources to reduce the resource supply system stresses of Houston; (iv) Benchmarking a solar energy supply system against a fossil fuel energy supply system; and (v) Assuming perfect foresight, how and at what cost can disruptive events be mitigated. Lessons learned from the previous project development include reflections on system definition, data acquisition, scale and assessment metrics.

References

[1] J. Cook, M. Di Martino, R. C. Allen, E. N. Pistikopoulos, A. Avraamidou. A decision-making framework for the optimal design of renewable energy systems under energy-water-land nexus considerations. Science of The Total Environment, 827, 2022, https://doi.org/10.1016/j.scitotenv.2022.154185

[2] M. Di Martino, S. Avraamidou, J. Cook, E. N. Pistikopoulos. An Optimization Framework for the Design of Reverse Osmosis Desalination Plants under Food-Energy-Water Nexus Considerations. Desalination, 503, 2021, https://doi.org/10.1016/j.desal.2021.114937

[3] B. Daher, R. H. Mohtar, S. H. Lee, A. Assi. Modeling the Water-Energy-Food Nexus: A 7-Question Guideline, Water-Energy-Food Nexus: Principles and Practices 229, 57, 2017, https://agupubs.onlinelibrary.wiley.com/doi/abs/10.1002/9781119243175.ch6

[4] B. Daher, R. H. Mohtar. Water-energy-food (WEF) Nexus Tool 2.0: guiding integrative resource planning and decision-making, Water International, 40, 5-6, 2015, https://doi.org/10.1080/02508060.2015.1074148

577e Virtual Experimentation Modelling Direct Air Capture of Carbon Dioxide

Joseph Menicucci, Raguram Thiagarajan, Srinivas Rangarajan

The use of interactive visualizations to convey complex concepts offers multiple advantages. First, students engage in active learning of the material, which leads to greater concept retention than with the traditional classroom. Second, students can learn at "real time" the effect of changes to parameters, thereby understanding the fundamental concepts via inquiry that is facilitated by an instructor; that is, such modules serve as hands-on "computational" experiments. Third, since these simulations are easier to perform than experiments, one can literally run hundreds of settings to relate inputs and parameters of a problem with its output through real time visualization.

We have implemented numerous data-driven models demonstrating our approach. Our current project models direct air capture of carbon dioxide via adsorption in a packed bed. Fundamental and applied relationships from reaction kinetics, thermodynamics, and mass transfer inform the model.

This paper will introduce a model for the direct air capture of carbon dioxide, outline the relationships informing the (virtual) experiment, demonstrate the model, and offer supporting materials for classroom implementation.

577f A Novel Inquiry Based Carbon Capture Laboratory Module

Iram Rahman, Amanda Simson, Benjamin Davis

Our currently unsustainable global carbon pollution epidemic is complex and requires innovative solutions and student understanding of its causes and effects. Inquiry-based learning (IBL) can promote a hands-on and individualized approach to education that encourages students to investigate environmental solutions that are meaningful to them. Through activities that require critical thinking, problem solving, and collaboration, students can develop the skills needed to address the issue of global GHG pollution by learning about how carbon dioxide can be removed from the atmosphere.

This paper is on the assessment of a carbon capture module introduced into the fall semester of a senior chemical engineering laboratory course. The module is meant to educate students about climate change and the emerging industry of carbon dioxide removal. Students built on progress made by previous groups, with initial groups starting with chemistry experiments and final groups ending with manufacturing and testing a direct air capture reactor with a 0.5L packed bed volume using aqueous NaOH absorbent. They assisted in creating a prototype reactor which was later characterized and modeled. A pilot reactor with an approximate packed bed volume of 13L is being developed to validate that the model can predict the performance of the scaled-up pilot reactor as part of a master's level project assessing the sustainability of mass-produced and distributed direct air capture compared to a traditional centralized plant.

In addition to teaching about an emerging industry, this curricular modification aimed to have students apply fundamentals of reactor design and have them collaborate with their peers to scale-up a chemical process. We created assignments to encourage collaboration, creativity, hands-on design, and construction. Evaluation of the IBL structure and climate change content was done by a mixed qualitative and quantitative survey, assessing aspects like the values of student-led modules, climate change relevant content, and working with prior groups to scale-up the process. This module will be used to develop understanding of alternate technologies as well; our vision is that students can contribute to a repository of student-led experimentation or follow new paths in emerging fields. We aim to demonstrate the value of supplementing traditional experiments with inquiry-based learning and of including climate change content in the primary chemical engineering curriculum.

577g The Case to Make Electrochemical Engineering Concepts Part of the Core Chemical Engineering Curriculum

Christopher Arges

Decarbonizing the global economy over the next few decades is arguably the greatest engineering challenge of our time. Numerous governments and corporations are racing to deploy green energy technologies and cleaner manufacturing processes to mitigate a climate catastrophe. Electrochemical engineering is central to the green technologies that are on the horizon. Although a very mature subject with a strong connection to chemical engineering content, the vast majority of undergraduate students in the United States get no formal exposure to electrochemical science and engineering unless they take a technical elective or pursue an advance degree on the topic. Rather than introducing a core electrochemical engineering course into the core chemical engineering curriculum, it is more effective to introduce small portions of the core tenants of electrochemical engineering content in each of the chemical engineering core courses. This talk provides a vision as to how each core chemical engineering course can spend 1 to 2 weeks a semester covering electrochemical engineering topics in a given chemical engineering course (e.g., thermodynamics, transport phenomena, process design, reaction engineering, process control, and unit operations laboratory). If this vision were adopted by all Chemical Engineering Departments in the United States, we would have about 10,000 graduating students per year produced that are proficient in electrochemical processes and able to meet the technological sea change of our time.

577h Reimagining Previous Design Projects for Energy and the Environment

Joseph Shaeiwitz, Steve Duke

Two previously published design projects [1] were modified to include elements related to energy and the environment and have been used successfully in class. Descriptions of the projects and assessment of the student experience with them will be presented.

One design project involves the process for producing ethylene oxide from ethylene and pure oxygen in the standard, highly exothermic partial oxidation reaction. Due to the subsequent combustion of ethylene oxide, the process is typically run at low single-pass conversion, with methane circulating to provide thermal ballast. Keeping the temperature low maintains satisfactory selectivity for the desired product. Current research and technological advancement provide a novel alternative. The technology exists to crush solid catalysts and put them in a solid mesh in a packed bed. The advantages are reduced mass transfer resistances and lower pressure drop, since there is more void space. If the mesh is made of conducting material, heat can be removed faster in the reactor. [2-5]

This project was used in the 2019-2020 academic year. Students were required to simulate the impact on reactor performance of increasing the heat transfer coefficient. All observed that higher conversion could be achieved without a high selectivity. As a design project, there are many opportunities for students to evaluate alternatives. These alternatives include using air vs. pure oxygen feed, operating the reactor above the UFL or below the LFL, and the choice of reactor operating conditions, including single-pass conversion. Therefore, this project provides evaluation of alternatives and multiple, realistic constraints. The choice to operate at higher single-pass conversion requires less recycle, thereby reducing the reheating and compression load, resulting in reduced energy usage and smaller equipment, which is a form of process intensification.

The second project involves the synthesis of dimethyl ether (DME) from methanol, which is also available (book). The reaction is reversible, $2CH_3OH\ CH_3OCH_3 + H_2O$. A typical process involves pure methanol as the feed, where methanol is produced from syngas by $CO + 2H_2$ ® CH_3OH. There is research underway [6] to produce methanol from $CO_2 + 3H_2$ ® $CH_3OH + H_2O$. The carbon dioxide could be recovered from power plants, for example. Therefore, this project clearly provides environmental and social considerations. The project used during the 2022-2023 academic year involved an equimolar methanol and water feed. For simplicity, the pure methanol and equimolar methanol-water feeds were used, thereby avoiding the impact on the water-gas shift equilibrium on either methanol synthesis reaction. To be clear, methanol synthesis was not included in the design project, though it could be, since equilibria and kinetics are known. [7] The opportunity for students could be to compare both process on an economic basis. For the equimolar methanol-water feed, one significant consideration for the students is whether to purify the methanol first. A critical decision is how pure the methanol needs to be as it enters the reactor. Choosing not to purify the methanol drives the DME equilibrium to the left, but the effect of the amount of water in the feed needs to be evaluated. These considerations are complex and reinforce students' understanding of both reaction kinetics and reaction equilibrium. Having students compare the economics of the pure methanol feed case and the methanol-water feed case offers experience with the cost/benefit trade-off of what appears to be an environmentally beneficial process compared to a traditional process. This project is in the process of being used for the 2022-2023 academic year, so summary of student results is not yet available but will be included in the presentation.

Process assignments and non-optimized Aspen Plus simulations for these design projects will be made available by the authors.

1. Turton, R., J. A. Shaeiwitz, D. Bhattacharyya, and W. B. Whiting, *Analysis, Synthesis, and Design of Chemical Processes, 5th ed,* Pearson, Boston, 2018, Appendix B.

2. Tatarchuk, B.J., *Preparation of Mixed Fiber Composite Structures*, U.S. Patent 5,304,3301994

3. Murrell, L. L., F. M. Dautzenberg, R. A. Overbeek, and B. J. Tatarchuk, U.S. Patent Application 2002/0068026A1, 2002.

4. Sheng, M., H. Yang, D. R. Cahela, B. J. Tatarchuk, *J. Catalysis*, **281**, 254-262, 2011.

5. C. B. Mukta, N. R. Rayaprolu, S. Cremaschi, M. R. Eden, and B. J. Tatarchuk, *Techno-Economic Study or Intensified Ethylene Oxide Production Using High Thermal Conductivity Microfibrous Entrapped Catalyst*, Proceedings of the 14th International Symposium on Process Systems Engineering (PSE 2021), Kyoto, Japan, 2022, http://dx.doi.org/10.1016/B978-0-323-85159-6.50116-0

6. Pérez-Fortes, M., J. C. Schöneberger, A. Boulamanti, and E. Tzimas, *Applied Energy*, **161**, 718-732, 2016.

7. Mevawala, C., Y. Jiang, and D. Bhattacharyya, *Applied Energy*, **204**, 163-180, 2017.

577i Promoting Resource Management in Teaching Laboratories: A Case Study for Effective Implementation of Sustainable Practices

Mohamed Maher, Benjamin Tan, Pavan Inguva, Jerry Heng, Umang V. Shah

Sustainability is top on the agenda for organizations with growing public demand to address carbon emissions. Over the last three decades, the UK has reduced its emissions by 44% and is the first G7 to legislate the target of achieving net-zero carbon emissions by 2050 [1]. In this respect, we address the need to quantify the environmental footprint of research and teaching laboratories. Although most may assume that the environmental impact of a laboratory may be low or insignificant, this is not necessarily the case, and one must consider the cumulative effect on the nation's ability to achieve its net-zero goal. We propose that educational and research institutes collaborate to advocate for environmental protection by adopting sustainable laboratory practices to reduce the carbon footprint.

A recent report by The Royal Society of Chemistry [2] states that laboratories are responsible for 60 – 65% of electricity consumption within research institutes, while some research laboratories produce large amounts of waste, which accounts for 1.8% of the world's plastic production. Therefore, improving the sustainability of individual laboratories is believed to be the key to achieving Imperial College London's target of becoming a sustainable and net-zero carbon institution by 2040. As Imperial College London gravitates toward a sustainable economic model [3], the UG teaching laboratory in the Department of Chemical Engineering at Imperial College London has taken the initiative to develop rigorous solutions that tackle the environmental impact of laboratories.

In educational institutes, there are typically two sources of carbon emissions: direct and indirect. The former is generated due to the sources owned and controlled by the educational institute's Estates department. On the other hand, the indirect emissions arise from activities across the educational institute and are not directly managed by the educational institute. Herein, we focus on the indirect emissions resulting from laboratory activities and contribute to the educational institute's carbon footprint.

Teaching laboratories are energy and resource-intensive by nature [4], and one must take a holistic approach to alleviate the environmental impact while maintaining the teaching activities' objectives. Over the past decade, several sustainable laboratory practices have been introduced, such as efficient waste management, minimizing water and energy usage, and utilizing eco-friendly materials. Nevertheless, it is crucial to view the entire lifecycle of the products being used and the processes carried out throughout the day, from procurement to disposal, to reduce the environmental impact. Implementing sustainable lab practices has been an ongoing effort at Imperial College London over the past few years as it benefits the environment and results in economic and social advantages. Imperial College has proactively implemented the Laboratory Efficiency Assessment Framework (LEAF), leading to 6 silver and gold certificate awards [4].

We developed and implemented an innovative-multifaceted and comprehensive strategy to achieve sustainability, intending to achieve net-zero emissions by 2026. The laboratory accommodates 22 experimental facilities at pilot and bench scales, serving three core undergraduate modules. These modules run in parallel from October – December and January – March, which requires the laboratory to operate for 6 hours per day for the complete module duration. Furthermore, the laboratory is equipped with 15 analytical equipment, *i.e.*, HPLCs, GC, particle size analyzers, DSC, Spectrometers, density meters, etc., that provide the characterization facilitates for various teaching lab projects. The laboratory also has auxiliary equipment such as fume cupboards, lab freezers, dishwashers, weighing balances, hot and stirrer plates, and overhead stirrers. During the term time, we estimated the energy, water, and solvent consumption for the three modules and identified the measures that can be implemented to reduce the consumption.

Figure 1 Approach to deliver sustainable Cheme Teaching Labs

To address the concerns above, we implemented a three-step ameliorated approach (see Fig.1) to reduce the energy, water, and solvent consumption during these modules. First, to alleviate the energy consumption in the lab, a hybrid-microgrid system based on Photovoltaics and a Fuel cell system was designed and is currently being implemented. The aim is to generate 16 kW to power 4 pilot-scale experimental rigs, which will be taken off the Imperial College grid resulting in a marginal reduction in electrical consumption. A further reduction in electrical consumption has been achieved by eliminating the usage of individual bath heaters and alternatively utilizing a closed-loop heating system that serves the entire laboratory. We have implemented strict regimes around the hydraulic benches and other fluid mechanics-based projects, where all water is reused and recycled. No fresh tap water is used in the labs except for preparing some solutions (i.e. NaOH solution) and cleaning glassware.

The water consumption within the lab has been optimized by utilizing closed loop cooling and heating water systems, replacing traditional methods, *i.e.*, tap water as a cooling medium in fume cupboard condensers and individual temperature control systems requiring tap water as a cooling medium, resulting in a significant reduction in water consumption. A pilot-scale ion exchange facility was also commissioned and operated to produce demineralized water for lab operations, resulting in some lab wastewater being recycled.

Finally, we introduce an experimental ecosystem to recirculate the solvents and significantly reduce solvent purchases. The ecosystem is schematically demonstrated in Figure: 2. This will help us substantially reduce fresh solvent usage from non-sustainable sources while achieving the learning outcome of practical education in an authentic setting. We are currently working on establishing collaborations, allowing us to obtain other organic and inorganic chemicals used in the facilities using green chemistry routes.

Figure 2 Experimental ecosystem to recirculate and reuse solvents within the teaching facilities.

Achieving sustainability is paramount in teaching laboratories as our nations work toward a sustainable future. While implementing sustainable measures and delivering a seamless learning experience may appear challenging, a balance between sustainability and upholding the learning objectives can be achieved through reasonable adjustments to the workflow. The case study presents a novel framework to achieve sustainability in a teaching-oriented laboratory, which has not been investigated thoroughly, and provides insight into the significance of addressing indirect emissions. The results of incorporating three key sustainable practices into teaching laboratory operations show the potential for not only significant reductions in energy, water, and solvent consumption but also led to the enhancement of the learning outcomes. Therefore, educational institutes must take further strides toward tackling indirect emissions and gradually scale up efforts to achieve collective sustainability goals.

References

[1] Department for Energy Security and Net Zero (2022) Net zero strategy: Build back greener, GOV.UK. GOV.UK. Available at: https://www.gov.uk/government/publications/net-zero-strategy (Accessed: April 2, 2023).

[2] 'Sustainable Laboratories - Practices to Reduce the Environmental Effects of Research.' n.d. Royal Society of Chemistry. Accessed April 2, 2023. https://www.rsc.org/new-perspectives/sustainability/sustainable-laboratories/.

[3] Sustainable imperial (2021) Imperial College London. Imperial College London. Available at: https://www.imperial.ac.uk/sustainable-imperial/ (Accessed: April 2, 2023).

[4] Leaf - a new approach to achieving laboratory sustainability (no date) LEAF - A New Approach to Achieving Laboratory Sustainability | Sustainability Exchange. EAUC. Available at: https://www.sustainabilityexchange.ac.uk/leaf_a_new_approach_to_achieving_laboratory_sus (Accessed: April 2, 2023).

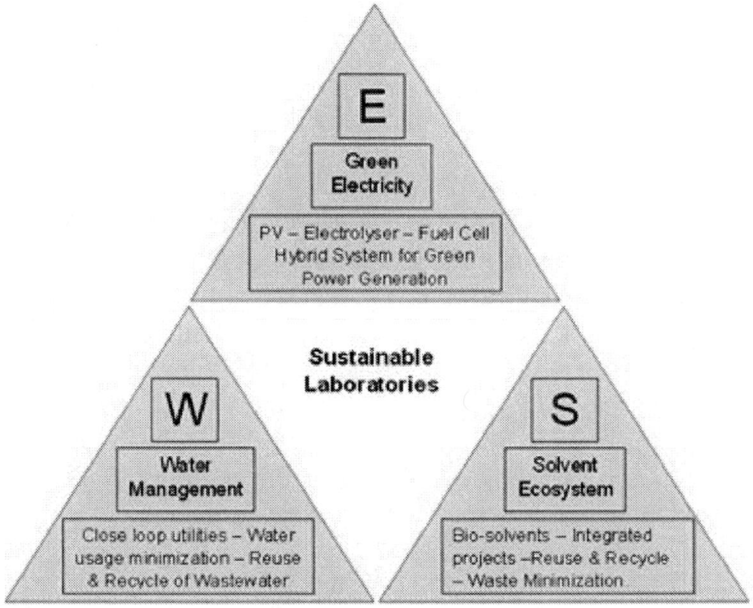
E
Green Electricity
PV – Electrolyser – Fuel Cell Hybrid System for Green Power Generation
Sustainable Laboratories
W
Water Management
Close loop utilities – Water usage minimization – Reuse & Recycle of Wastewater
S
Solvent Ecosystem
Bio-solvents – Integrated projects –Reuse & Recycle – Waste Minimization

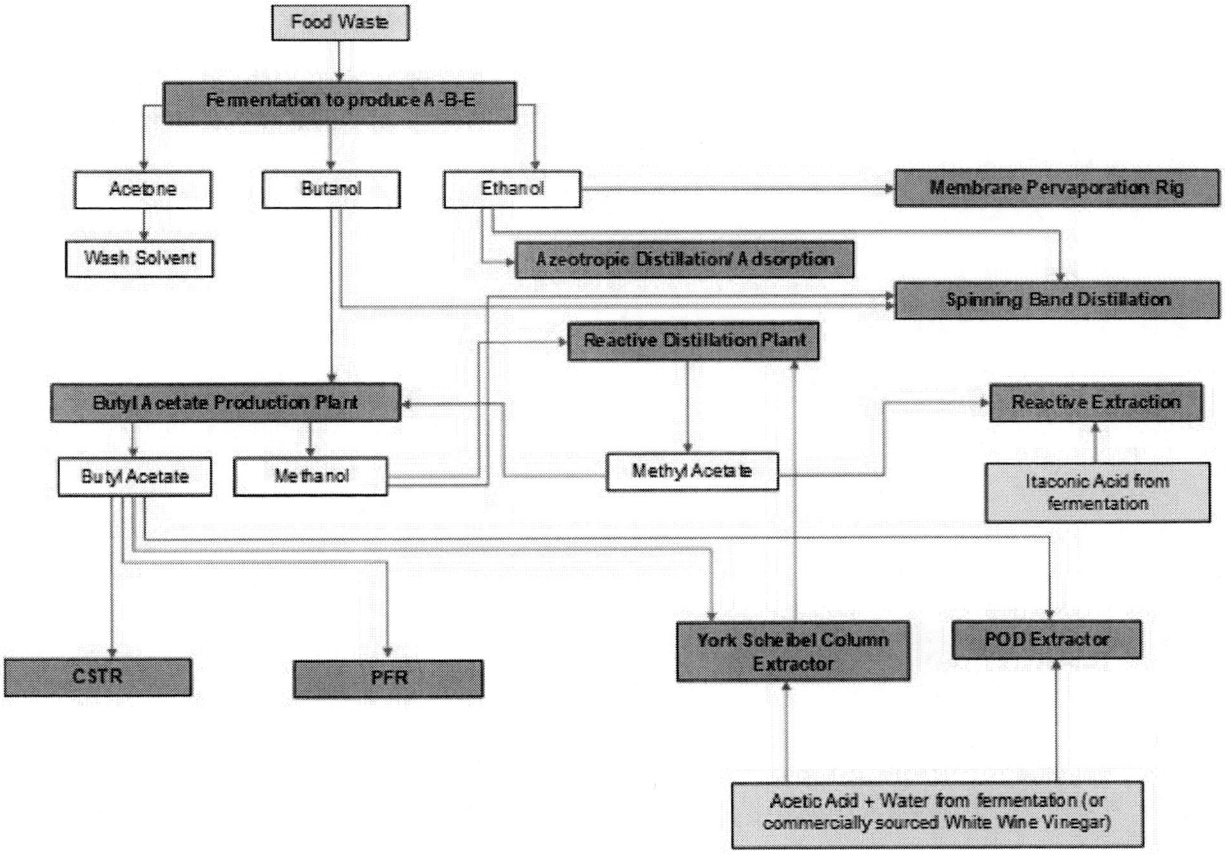
Food Waste
Fermentation to produce A-B-E
Acetone
Butanol
Ethanol
Membrane Pervaporation Rig
Wash Solvent
Azeotropic Distillation/ Adsorption
Spinning Band Distillation
Reactive Distillation Plant
Butyl Acetate Production Plant
Reactive Extraction
Butyl Acetate
Methanol
Methyl Acetate
Itaconic Acid from fermentation
CSTR
PFR
York Scheibel Column Extractor
POD Extractor
Acetic Acid + Water from fermentation (or commercially sourced White Wine Vinegar)

445a Would Students Prefer to Watch Lecture Recordings Rather Than Attend Class: A Follow-up Study

Matthew Cooper, E. Daniel Cárdenas-Vásquez, Aaron Frye

Many ChE instructors noticed changes in classroom attendance trends once classes fully returned to in-person delivery upon easing of the COVID-19 pandemic; anecdotally, classes which previously had attendance rates of nearly 100% dropped below 50%. Prior work presented at the 2023 ASEE Conference analyzed open-ended student responses to survey questions using thematic qualitative coding in order to better understand the perceptions of senior undergraduate ChE students in Fall 2022 on factors influencing their classroom attendance and engagement. In that work it was indicated that some students consider watching lecture recordings to be a suitable replacement for in-person attendance (or they even prefer online learning), and that the lack of an attendance policy for the studied course played an important role in students deciding not to attend class in person.

This work provides follow-up analysis regarding application of the findings of this previous work to another course in the undergraduate ChE curriculum. Policies regarding lecture recordings and in-person classroom attendance were adjusted based on findings of the previous work and applied in a core ChE course (numerical methods) in the sophomore year. It was observed that in-person attendance rates exceeded 90%. In order to more deeply understand the factors which encouraged students to attend class, a survey question was included on end-of-semester course evaluations which asked students whether they would have regularly attended class in person if there was no attendance policy but lecture recordings were still available for asynchronous viewing. Student responses to this question were qualitatively coded using a thematic analysis method similar to that employed in the prior work; results from this analysis will be presented.

445b Epistemic Metacognitive Reflection on Concept Questions in Active Learning

Lorena Grundy, Milo Koretsky

Many engineering problems assigned in undergraduate classes are numerical and are solved using equations and algorithms. Concept questions, which can be administered in class using active learning pedagogies, aid in the development of conceptual understanding as opposed to the procedural skill often emphasized in numerical problems. The effectiveness of concept questions is often assessed based on the correctness of students' answers. In the literature, student thinking has been probed via their explanations of their answers, and metacognition—students' thinking about their own thinking—has been investigated based on their self-reported confidence. In this work, we are interested in students' *epistemic* metacognition, or their thinking about their own knowledge and learning, in response to concept questions. We administer a concept question to over 300 students in six courses taught at six diverse institutions and find no statistically significant differences in answer correctness or confidence between the classes. We assess epistemic metacognition by asking students whether the question made them think deeply about the course material, and why. Across the classes, students consistently report that they are not accustomed to such non-numerical concept questions, but they grapple in different ways with the experience. Some students frame engineering as inherently numerical, and thus do not value the conceptual understanding assessed by the question, while others recognize that developing conceptual knowledge is useful and will translate to their future engineering work. We conclude that groups of students with similar levels of conceptual understanding may differ in another dimension of learning: their epistemological framing of engineering problems. We discuss the ways in which the differences in teaching methods used by the six instructors in this study may have contributed to the differences in their students' epistemic metacognition.

445c Beyond the Lab: Virtual Exchanges for Chemical Engineering Laboratories

Fernando Mérida Figueróa, Patricio Zapata

Chemical engineering laboratories are an integral part of undergraduate chemical engineering education, providing students with hands-on experience in the practical application of theoretical concepts. These labs are designed to primarily simulate industrial processes, allowing students to develop a range of essential skills, including experimental design, data analysis, and problem-solving. They also offer unique opportunities for students to work collaboratively, improving their communication, teamwork, and leadership skills. While these labs focus on technical skills, they may not provide opportunities for students to engage in supplementary activities beyond the lab that further enhance their experiential learning. Examples include opportunities for students to learn about or appreciate different cultures, development of foreign language proficiency, and student-centered experiment discussion with peers from a different institution. This can be important for developing a global mindset in today's interconnected world.

Chemical engineering processes are increasingly global in nature, and working with diverse cultures and languages is a common occurrence in industry and graduate school pathways. Thus, students who develop multicultural competencies, foreign language proficiency, and improved knowledge of location-specific experimentation during their education are better prepared to succeed in the profession. As such, it is important for institutions to balance the technical focus of chemical engineering teaching labs with opportunities for students to engage in broader learning experiences.

We propose the design and subsequent integration of a Virtual Exchange (VE) as an alternative to enhance the student learning outcomes in chemical engineering laboratories. VEs offer a unique opportunity for students to engage with peers from different countries and cultures without leaving the classroom, thus broadening their exposure to diverse perspectives and practices. In this way, VEs could help bridge the gap between the typical lab-based class structure and real-world experience by simulating the challenges and opportunities of global collaboration. Furthermore, virtual exchanges could be tailored to incorporate language and cultural competency development, as well as industry-specific experimentation knowledge. However, it is essential to ensure that virtual exchanges are well-structured, well-resourced, and well-supported to achieve the intended learning outcomes including those of societal impact in a global context.

The Chemical Engineering Departments at the University of Florida (USA) and Universidad de Concepción (Chile) have established a partnership to develop a Virtual Exchange (VE) for their junior lab. The primary objective of this VE is to enhance the experiential learning of students by integrating discussions of experiments within a multicultural framework, coupled with technical and non-technical language exchange. These activities will be linked to the established student learning outcomes of each class, and instruments will be developed to evaluate the impact of VE-related activities on achieving these outcomes. The partner institutions have completed a mandatory five-week VE training program as the first phase of the project. The VE design process is currently in progress, with a focus on identifying synchronous and asynchronous activities, establishing timelines, determining assessment methods, and allocating VE-related activities within the course schedule for each institution

Overall, the use of virtual exchanges in chemical engineering laboratories has great potential to enrich students' learning experiences, increase their global awareness, and enhance their employability in a rapidly changing world.

445d Teaching with Cases in Chemical Engineering

Christopher V. H-H. Chen

More and more students are graduating from chemical engineering programs and heading into careers beyond traditional chemical engineering (i.e. commodity chemical). As such, there is a growing need for curricula and faculty to better prepare students to apply their problem solving skills to a wider variety of problems and fields, and expose students to the possibilities of chemical engineering training beyond the plant or refinery. Drawing on a wider range of examples allows for greater student motivation, an important for academic success in chemical engineering. Thus, a major driver for improving our teaching was to consider how we can help students see chemical engineering thinking as a more transferable and useful skill beyond the classroom, a known challenge when using traditional chemical engineering problems. This is not a new idea in chemical engineering education, having been advocated in previous calls for curricular change from the turn of the century, and continued innovation in effectively implementing pedagogical strategies to aid in this transfer of chemical engineering skills are still being sought.

Case-based instruction is a pedagogical approach common in law, medicine, and business, but is relatively less common in science and engineering. The National Center for Case Study Teaching in Science (NCCSTS) at the University of Buffalo promotes STEM case instruction, and maintains the Case Collection–peer-reviewed case database–with the National Science Teaching Association. Many of the sciences have a large number of cases, with 100+ chemistry entries in the collection's of 1000+ cases, for example. However, the NCCSTS Case Collection houses a paltry 13 chemical engineering cases as of Spring 2023, only 5 of which are appropriate at the introductory level when students are just starting to engage with the discipline (chemistry in comparison, has about 100 cases available through that database). Other resources for chemical engineering pedagogical resources (e.g., LearnChemE and AIChE Concept Warehouse) host few cases. The limited availability of chemical engineering cases shows the benefit to our field in the development of relevant case studies, especially at the introductory level where student experience is highly correlated with persistence and motivation within the major.

Since Fall 2019, we (the instructors authoring this paper) have begun implementing case-based learning within our Material & Energy Balances course (MEB, a.k.a. Introduction to Chemical Engineering) to help students focus on transfer of their burgeoning chemical engineering skills. In our MEB course, cases are implemented at a variety of time scales; cover a variety of sectors (e.g., commodity chemical, energy and the environment, biopharmaceutical); and place students as chemical engineers in a variety of roles (both traditional and non-traditional). The shortest cases are run as in-class activities (~50 minutes) where students are placed in more traditional chemical engineering contexts (e.g., process engineering), and tasked with making a process design decision. Students are guided during the following discussion to find contexts in which their choices may differ, even with the same calculated results. Examples include selecting between synthetic pathways that differ in production and safety, and improving the efficiency of chemical reactors and furnaces.

An intermediary scale for the cases are assigned as team homework assignments. During these cases, students are placed in roles beyond what may be seen as traditional chemical engineering jobs to address open-ended questions based on current events. Teams of 4-5 students are given two weeks to draft a short slide deck (5-10 slides) with their proposals that are later peer reviewed. Group case homework include pitching chemicals targets that use carbon dioxide and excess solar power in California as an entrepreneur; and advising congress on incentivizing alternative fuels as an AAAS fellow.

Finally, the longest time scale for cases in the course are assigned as final projects. During the last month, student teams (4-5) are tasked with advising on a case based on unproven technologies in a written report that draws on concepts from throughout the class. Selecting contemporary challenges assures that right answers are unknown, and allows students to focus on their argumentation using chemical engineering approaches over "correctness." Previous final projects in our class include:

- Evaluation of a pyrolysis process for carbon sequestration in partnership with Charm Industrial
- Analysis of Proton Technology's Blue Hydrogen Production from spent oil wells
- Metalysis of regolith (moon rocks) into aluminum and oxygen using the FFC Cambridge process as described in the novel *Artemis* by Andy Weir

We have also written cases for graduate level courses based on chemical startups to consider how they may apply their chemical engineering decision making skills in an area that is becoming more popular for our students to enter after graduating. These cases are written to help place students in a context similar to that of a guest speaker visiting the course, working both as an in-class case (~30 minutes of discussion), but also as a background document for the speaker. When the speaker comes to share their experience working in chemical engineering practice, they also work with the students to unpack the case. This helps students better prepare for when the guest comes to class, but also gives them experience making difficult decisions (e.g., how to select a chemical product to pursue, how to find customers and a target market) before they would have to do so for real in their future roles.

In this presentation, we share how we have implemented case-based learning in our MEB course and a graduate elective, and provide data to show the effect of implementation on student learning and course experience–both in the first iteration of MEB with cases, and when the use of cases has become more mature. In MEB, we found that our students still make progress towards technical learning objectives in the course when we displaced lecture and other homework assignments with cases at different scales as measured through exams. Through surveys, we found that implementing cases had the added benefit of increased self-reported progress towards the ABET 1-7 objectives and evaluations of the course as compared to MEB prior to case-based activities. Free response data in these surveys were coded and demonstrated that students recognized the real world applications of chemical engineering in these cases, and appreciated how the cases allowed them to apply their chemical engineering thinking to a variety of contexts.

Our intent in sharing is not to convince the audience that our approach is the way to integrate case-based activities into the chemical engineering classroom, but instead of help inspire and convince more of our colleagues to try using cases as another tool to engage students in learning chemical engineering. We plan to continue to write and share our developed cases, and to further explore how these cases can be used in other parts of the chemical engineering curriculum.

445e Teaching Process Calculations to Undergraduate Chemical Engineers: Changing the Evaluations Paradigm

Preeti Aghalayam

Introduction

Process Calculations, or Principles and Calculations in Chemical Engineering, is a required first year Chemical Engineering undergraduate course. Class sizes at all major universities tend to be large, and as it is a first foray into the field for the students, gaining conceptual clarity and understanding problem solving methodologies is a challenge. Furthermore, the voluminous amount of data required for solving problems, and the myriad sources of such data that have to accessed and chosen with careful discrimination, are problems faced by students.

As instructors, evaluations pose big hurdles. With large classes, and a just-out-of-high-school mentality of students (which means that they struggle to accept the open-ended nature of most engineering problems), crafting intelligent exams that enhance learning is difficult. In this paper, we propose a way of integrating different types of exams & evaluations into the design of the course. We highlight a means for using games (simple, easy to create ones) as both learning and evaluation tools in the semester-long course.

Overall, the purpose of a core course like Process Calculations is to provide students with significant advancement in their learning, so that they access the further courses in the semester comfortably and with enthusiasm. But as instructors the human aspects of managing student stress and ensuring that the evaluations are meaningful and a natural extension of classes, is important.

Philosophy of Evaluation Scheme

In our design of the course, we wanted to think of exams and evaluations as additional opportunities for student learning, and not merely a demonstration of the knowledge gained during the classes. Moreover, we also operated under the assumption that *effort* and *accuracy* are both important and students deserve marks (or points) , which count towards the final grade, for putting in measurable effort, as well as for getting to the correct answer. Finally, we wanted the evaluations to be fun (as possible).

Administrative and other Constraints

In our system, there is a requirement that core courses follow a common pattern of exams. Principally, they need to involve a "Big Final" which contributes at least 40% of the final marks leading to a grade. It is highly recommended that there are at least two other in-person, proctored exams (called quizzes) spread evenly across the semester. Within these, significant freedom is provided to instructors to plan the semester, as long as the plan is known to the students well in advance, preferably at the beginning of the term. The oversight is from a committee consisting of handful of other faculty from the department – student representatives are given the chance to articulate the concerns of the class during the meeting with the committee (held twice in the semester). A relative grading pattern is typical in our core courses.

Proposed Activities of Evaluation

Keeping within the spirit of the requirements at IIT Madras, we used a scheme of evaluation that involved the activities shown in Figure 1. The combination of these diverse means of evaluation provide the opportunity to implement the three ideological principles we set out for ourselves (focus on learning; reward effort; have fun), all the while meeting the administrative needs.

Of these, a part of the Big Final and one of the in-semester Quizzes were done in the conventional/traditional mode (but were always open book). All the other parts of the evaluation were specially designed so as to reward student effort significantly. Of course the games that were used in the class and for the evaluations, added a fun element (though they are also competitive in a way, with time pressure and so on).

During the semester, we developed a number of games that were particularly enjoyed by the students. One class of games was just to help students learn the course material better, and avoid having to memorise terms in a rote manner. For e.g. we designed a simple game on https://wordwall.net/, to identify the different parts of a flow chart by dragging and dropping pins. Readers can attempt to play this game here: https://wordwall.net/resource/28995785

We used anagrams, crosswords, match the following and several other simple motifs – students played these at their own time/pace and typically went back again to make sure they got all answers right. The other kind of games involved time pressure and were of the multiple choice kind, on platforms like kahoot! (https://kahoot.com/) or quizzizz (https://quizizz.com/admin). These were played in class, with lot of noise (including background music!) and a competitive spirit among students (a leader board was displayed after each question). The back-end reports were used for individual grading.

Of the other elements that we used in evaluation, one requires particular mention –the reflections form that students filled every week. We used an online forum for this, and the feedback was meant to be non-anonymous (parenthetically, we collected a lot of quick anonymous feedback via polling softwares like slido and mentimeter, in class, this was useful but in a different way). The reflection forms were opportunities for students to briefly talk about their learning during the week, mention their favourite topic and raise any concerns they had. Every student who filled a form was awarded a point towards the final grade. This is a type of evaluation where effort is rewarded, with no further analysis of accuracy. We found this useful – we obtained the feedback we needed, students felt enthusiastic about writing to us, and it helped in managing stress levels.

Conclusions

In recent years, we have shaken ourselves out of the long-held belief that a noisy classroom is bad - students are not paying attention the teacher and are not learning. We embraced collaboration and high decibel levels in our classroom, spent time providing students with many activities that they enjoyed (far more than our lectures, perhaps), and stood evaluation on its head. Yes, this made our classes very loud at times but there were giggles and smiles, and the student testimonials at the end, were nothing but encouraging. In terms of the performance of the students, the class average marks and the distribution of grades were well within reason. We have also made a big push to use technology appropriately in the classroom, to enable us to engage with students better. For the conventional exam components of the course, we are now working with partners to provide AI-enabled unique exam papers with auto-grading possibility, for each student. This will be experimented on later this semester. In this presentation, we look forward to illustrating our methods and discussing ways in which chemical engineering faculty can leverage them, particularly in large, core course classes.

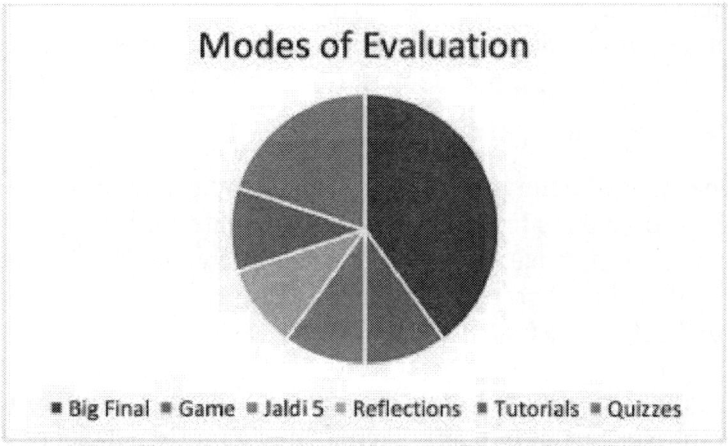

Figure 1 Different modes of evaluation used in Process Calculations at IIT Madras

445f Student Goal Orientation and Structure in a Problem-Solving Course

Carl Lund

According to achievement goal theory [1], students in a course can perceive four goal structures. When performing a task in that course they can adopt one of four corresponding goal orientations. Their perception of the course goal structure can influence the goal orientation they adopt. The four types of goals are mastery approach, mastery avoidance, performance approach and performance avoidance. Student adoption of a mastery approach goal orientation has been associated with adaptive learning strategies. For this reason, instructors attempt to create a mastery approach goal structure in their courses.

For present purposes, a problem-solving course is one where the instructor presents theory, derives equations, and shows how to use the theory and equations to solve problems. The learning objectives then are for students to be able to use the theory and equations to solve similar problems. In this study, the instructor in a problem-solving chemical engineering course changed aspects of the course delivery, particularly the manner in which homework was administered, in attempt to create a mastery approach course goal structure. These changes occurred in a required, junior-level course on kinetics and reaction engineering.

A particular focus of the changes related to homework problems. A student holding a mastery approach goal orientation seeks to know and understand how to solve each assigned problem. In contrast the goal of a student with a performance approach orientation is to get the correct answer for each assigned problem. The latter orientation can lead to strategies wherein the correct answer is obtained without substantially developing the ability to solve the problems independently.

Two more drastic changes in the course delivery were associated with homework assignments. First, homework assignments were graded exclusively on the basis of the effort put forth toward solving the problem irrespective of whether a correct answer was obtained. Second, the solution to each homework assignment was given to the students at the time the problem was assigned. Each problem statement said that the purpose of the assignment was for the student to gauge their ability to solve the problem under exam-like circumstances. It instructed the students to try to solve the problem referring only to an equation summary that would also be available when they were taking exams. It suggested that while solving the problem, they should only consult the provided solution if they became stuck and could not make progress. It encouraged them to compare their solution to the one that was provided after they had completed the solution with the intention identifying any lack of understanding or misconceptions they might hold so that they could be addressed.

The study sought answers to two questions: What goal structure and goal orientation did students perceive/hold at the start of this course? Did students' perception of the course goal structure or the goal orientation they adopted change as the course progressed? The approach involved administering three anonymous surveys. One was administered at the start of the course, one at the middle of the 15 week semester, and one at the end of the course.

The survey questions were taken, with very minor modification, from the Patterns of Adaptive Learning Survey (PALS). It consisted of 28 PALS items and one original item. The PALS items have been used in many prior studies. The form scales to measure three of the four (all but mastery avoidance) types of goal orientations and goal structures held or perceived by the respondents. Summary statistics were generated and a confirmatory factor analysis (CFA) was performed for each of the individual surveys.

The surveys indicate that at the start of the course approximately twice as many students perceived a mastery course goal structure and held a mastery goal orientation as did performance approach goal structure and orientation. There was very little change across the three surveys, but the number of responses was small (43, 40 and 34 for the start, middle, and end of semester surveys, respectively). Because the surveys were anonymous, it was not possible to examine changes at the individual student level. The CFA analysis did not yield models with a good fit.

This work was preliminary, and the results were unexpected. A number of changes are planned before the next offering of the course.

[1] Bardach, Lisa, Sophie Oczlon, Jakob Pietschnig, and Marko Lüftenegger. 'Has Achievement Goal Theory Been Right? A Meta-Analysis of the Relation between Goal Structures and Personal Achievement Goals.' _Journal of Educational Psychology_ 112, no. 6 (2020): 1197--1220.

[2] Midgley, C., Maehr, M. L., Hruda, L. Z., Anderman, E., Anderman, L., Freeman, K. E., Gheen, M., et al. 'Manual for the Patterns of Adaptive Learning Scales.' University of Michigan, 2000.

445g Teaching Degree-of-Freedom Analyses to MEB Students with 'Specify the Specifications' Problems.

Vincent Tocco Jr.

Performing a degree-of-freedom analysis is an essential problem-solving step for Chemical Engineers in *Material and Energy Balances* and beyond. Before deriving and solving equations, the problem-solver must first ensure that the number of constraints matches the number of unknowns— the problem has multiple contradictory solutions if there are too many constraints, but the problem is not solvable if there are too few constraints. Despite this importance, novice Chemical Engineering students sometimes skip this step and assume (sometimes incorrectly!) that any problem assigned for coursework must be appropriately specified.

To encourage students to learn, use, and practice this essential skill, an original genre of *Material and Energy Balance* problems are presented here in which many potential constraints are presented in list form, as opposed to being explicitly given in the problem statement. Students are instructed to accept these potential constraints as true only until the quota of constraints (as defined by the number of unknowns from the degree-of-freedom analysis) has been met. They must disregard all further constraints as superfluous.

Due to the uncertainty about which information should be accepted as true, these type of "specify your specifications" problems can be quite stressful for students lacking in self-confidence. Hence, these problems do not work well for high-stakes individual assessments, such as exams. However, they work very well for in-class problem-solving sessions in small groups, where spirited debate among peers can help students to understand the importance of a degree-of-freedom analysis and gain confidence. These problems also work well for homework assignments, where students may experience the agony of attempting to solve an over- or under-specified problem without the time pressure of an exam.

445h Randomly Generated Assessments with Achievement-Based Grading in Chemical Engineering

Adam Ekenseair

Achievement-Based Grading (also similarly referred to as Specifications Grading or Mastery Grading) is an assessment structure where students work towards demonstration of learning and skills and a final grade is awarded based on the number and level of achievements successfully completed. While implementations are varied, common themes include distinction between simple and more complex assessments, significant flexibility in when (and often in what order) learning and skills can be demonstrated, and opportunities to grow through failure by re-taking unsuccessful assessments. Some other aspects that may be found in particular implementations include pathways to the highest grade available without having to show achievement in all skills covered in the course and the ability to demonstrate achievement though a variety of means.

This assessment structure has appealed to educators for a number of reasons. One is the potential to streamline grading and make delivery of challenging material more manageable, though this ultimately depends on the particular implementation. Some more generally applicable benefits are associated primarily with a shift in student perceptions of what assessment means and how progress through curricular topics is accomplished. The focus is placed on the topical material and course outcomes and students' ability to demonstrate those outcomes by the end of the term. Instead of worrying (or perhaps complaining) about how points were lost on one assessment or another, students instead have a path forward where their grade is earned based on demonstrated achievements and where they are encouraged to revisit topics until mastery can be shown. This structure also facilitates a straightforward means of directly challenging the average student to achieve a higher degree of learning in a course. Furthermore, instructors can readily project a supporting role in the progression of students' development. Instead of feeling like grades were unfairly or even arbitrarily determined, students can see the expectations from the beginning and relate their final grade to their own accomplishments.

A variation of Achievement-Based Grading has been successfully implemented in Chemical Engineering core courses, including Thermodynamics 1, Material and Energy Balances, and Separations and Plant Design. The details and evolution of this implementation, lessons learned, student feedback, and impacts on student learning will be discussed, with a particular emphasis on the incorporation of randomly generated problems deployed as Canvas quizzes for direct assessment of learning targets.

445i Correlations and Reproducibility in Reading Participation and Auto-Graded Homework Completion When Using an Interactive Textbook for Material and Energy Balances

Matthew Liberatore, Samantha Yanosko

Combining interactive reading participation with online homework generate large, real-time data that can help quantify student engagement and learning. A fully interactive textbook for a material and energy balances course measures over 1,400 reading interactions per student as well as the fraction correct and number of attempts on up to 700 auto-graded homework questions during a semester. In total, seven cohorts representing over 600 students and over 700,000 reading interactions exhibit interesting correlations between reading participation and final course grades. The interactive content applies some best practices from learning sciences, including using visuals in animations, chunking content into organized sections and subsections, immediate feedback when using multiple choice, matching, and true/false learning questions, and varying the reading activities. Since reading participation is an effort-based grade, the correlation between reading participation and final course grades may be considered surprising. Also, when allowing unlimited attempts before homework is due, students earned a median correct of 91% or higher across six cohorts on auto-graded problems. The randomized questions are also used by students as a means of practicing problem solving with thousands of attempts after correct being logged in recent cohorts. Students final grades more strongly correlated with their scores on auto-graded problems than reading participation. Statistically significantly lower competition of auto-graded problems was measured for each subsequently lower grade, i.e., B students answer statistically more auto-graded problems than C students. Overall, these correlations provide data to weigh the importance of formative and summative assessments in a material and energy balances course.

476a Trends in Teaching-Focused Faculty in Chemical Engineering

Stephanie Velegol, Taryn Bayles, Katie Cadwell, Lisa Bullard

Teaching-focused (non tenure track) faculty make up 15% - 25% of chemical engineering departments. These faculty not only teach core chemical engineering courses, they are also often the faculty in charge of ABET, serve as advisors both to students and to student groups, perform engineering educational research and lead the departments in implementing new teaching strategies. There have been recent strides to help support these faculty including creating pathways for promotion and funds to attend conferences. In addition, the educational division of AICHE has sponsored a virtual support group for chemical engineering teaching-focused faculty that meets monthly on zoom. This group of 60 faculty has discussed ways to get involved in AICHE and ASEE, ways to get involved in engineering educational research and ways to advocate for themselves at their individual institutions. This prompted the authors to create a survey to the members of the group (and beyond) in order to find ways that the faculty are currently being supported and to explore ways that this support can be extended. The purpose of this talk is to share the results on this survey and discuss ways that these faculty can be supported by their peers and institutions.

476b Stategies to Support Contingent Faculty in Engineering for Better Work Conditions

Homero Murzi, Marisela Martinez-Cola

This faculty development workshop will help participants better understand the barriers that contingent faculty in engineering face. Contingent faculty are non-tenure track faculty who had positions that are contract-term bound or are of a temporary appointment. Literature suggests that contingent faculty face a slew of challenges that their tenure-track colleagues don't such as job insecurity, inconsistent health care, low pay, lack of sense of belonging, and under-resourced working conditions. The lack of stable working conditions for these faculty, in turn, negatively impact student persistence and retention.

This workshop will present some findings from the authors' research following by reflection activities and dialogue on professional development opportunities to help faculty identify strategies to help them advocate for better working conditions. While the workshop is tailored to contingent faculty, graduate students, postdocs, administrators, and tenure-track faculty are welcome to join.

476c Community of Practice in Action: A Participatory Framework for Faculty Professional Development in Chemical Engineering Education

Sindia M. Rivera-Jimenez

In the past decade, there has been an increasing call for chemical engineering faculty to incorporate the social implications of techno-economic fundamentals into their undergraduate courses. However, achieving this requires significant changes in teaching, learning, and assessment methods. To address this, this work proposes a conceptual framework for faculty professional development in engineering education, using participatory action approaches and strategic agency. The Community of Practice in Action (CoPA) model consists of five components: community participation, reflection, strategy, action, and time, which fosters a culture of continuous learning and improvement within professional societies like AIChE. The CoPA model provides a flexible framework for professional development that can be adapted to the needs of individual faculty members and diverse institutional cultures.

Professional societies can play a unique and convenient role in supporting faculty transformation and, subsequently, their students because they are established communities of practice that disseminate shared understanding through evolving standards, norms, and perspectives. They can provide a safe and collaborative space for faculty to create critical consciousness to support their transformation. The CoPA model encourages faculty to reflect critically on their teaching practices, collaborate with colleagues, and implement evidence-based changes to improve student learning outcomes. By adopting the CoPA model, professional societies can support their faculty in addressing the complexities of teaching and learning in diverse and rapidly changing contexts while fostering a culture of continuous learning and improvement.

To demonstrate the effectiveness of the CoPA model, a cohort was created from September 2022 to May 2023, consisting of chemical engineering faculty from various institutions. The cohort participated in a CoPA to develop scholarly activities incorporating social responsibility themes into the chemical engineering curriculum. The cohort engaged in semi-structured activities that fostered collaboration, reflection, and action to promote social responsibility in engineering education. The presentation will highlight the successes and challenges of this cohort and the impact of the CoPA model on faculty professional development and curriculum transformation.

In summary, the proposed CoPA model provides a flexible and collaborative framework for faculty professional development in chemical engineering education. Professional societies like AIChE can play a crucial role in supporting faculty transformation, and the effectiveness of the CoPA model will be demonstrated through the cohort's experience. By fostering a culture of continuous learning and improvement, the CoPA model can help address the complexities of teaching and learning in diverse and rapidly changing contexts while promoting social responsibility in engineering education.

476d Change Theory: The Use of the Henderson Foursquare and the Ascn Change Dashboard

Donald Visco Jr., Lisa Bullard, Jason M. Keith, David Silverstein, Charles Henderson

The National Science Foundation requires proposals in certain areas to specify *how* the intervention the authors propose in their work will actually result in the desired change. In other words, the mechanism by which this change should occur. Such phrasing is typically known as the "theory of action" or "theory of change".

In this work, we look to provide a brief introduction to popular theories of change that have been part of the STEM literature for the past two decades. However, as noticed by Henderson and co-workers, not all change theories are valuable for all desired changes. Indeed, it is crucial to match an appropriate change strategy with the desired change. To that end, Henderson and co-workers suggest a foursquare that looks at change on two different dimensions: (1) is the change bottom-up, where the final change state is being developed [B] OR is the change top-down, where the final change state is known [T] and (2) is the change looking to impact individuals [I] or is the change looking to modify structures/environments [E]. Once a change can be viewed through this foursquare lens [B/I, B/E, T/I, or T/E], an appropriate change strategy can be selected that is most appropriate for the placement within the foursquare.

Operationalizing such changes can be facilitated through the Accelerating Systematic Change Network (ASCN) Change Dashboard. In this work, we will briefly introduce this Change Dashboard and how it might be used to implement a desired change, translating strategy to workable tactics through a group thought development exercise with with some examples relevant to chemical engineering academic departments and their curricula.

476e Making the Move: Navigating a Transition from Disciplinary Engineering into Engineering Education Research

Sindia M. Rivera-Jimenez, Sarah Wilson

The teaching and research interests of faculty often shift as they navigate their careers in higher education. Exposure to student experiences and challenges in the classroom can lead faculty towards interests in engineering education. While research ideas often come easily, knowing how to conduct rigorous research in engineering education can be more challenging. Navigating a transition into social science research methods can be difficult and intimidating, resulting in faculty choosing not to pursue these research interests. Therefore, this workshop is designed for faculty in engineering who are interested in starting research in engineering education. Through the workshop, we will provide a pathway for navigating the development of an engineering education research project. Faculty will have the opportunity to work through a scaffolded activity to identify a research area of interest and brainstorm potential research methods. By the end of the workshop, faculty should understand the steps necessary to establish a rigorous and ethically sound study in engineering education.

564a Engineering Student Mental Health and Help Seeking: Analysis of Data from the Healthy Minds Study

Matthew Whitwer, Sara Lamer, Brenna Gomer, Joseph H. Hammer, Sarah Wilson

Mental health concerns such as depression, anxiety, and suicidal ideation are becoming increasingly prevalent on college campuses, leading some to label higher education as suffering from a mental health crisis. Within engineering, the rigorous courseload and highly competitive nature of such programs creates a "culture of stress" which could be negatively impacting students' mental health. However, published works have not provided a clear comparison between engineering students and students from outside engineering on the prevalence of mental health concerns or the likelihood of seeking help for those concerns. Additionally, few studies have looked at how identity factors such as gender might impact the interpretation of these results across engineering and non-engineering student populations. This study used publicly available data from the 2019-2020 Healthy Minds Study to understand how self-reported mental health distress and professional help seeking differs between undergraduate engineering and non-engineering student populations, as well as how it varies between engineering student demographic groups.

The Healthy Minds Study is an annual survey conducted at universities across the United States that examines factors related to mental health in student populations. The dataset used in this study contained responses from over 56,000 students, with almost 8,000 students from engineering majors, and 48,000 students from non-engineering majors. Regression analyses were used to examine each mental health and help-seeking variable across demographic groups and fields of study (e.g., engineering, business, humanities). When comparing across all engineering and non-engineering students, engineering students reported symptoms associated with lower levels of both anxiety ($p < .001$) and depression ($p < .001$). However, when controlling for gender, both male and female engineering students reported statistically similar levels of both anxiety and depression as compared to non-engineering students. Among distressed undergraduate students, engineering students were significantly statistically more likely than several other fields to have seen a therapist or counselor in the prior twelve months, when controlling for covariates ($p<0.05$), and no fields of study were statistically significantly less likely than engineering to have. Results from this study identify important differences in mental health status and help-seeking behavior across engineering and non-engineering student populations and highlight the importance of further research identifying both the cause of this treatment gap, and ways to reduce it so that engineering students can live full, productive lives.

564b Implementation of a College-Wide Engineering Student Mental Health and Wellness Training

Sarah Wilson, Joseph H. Hammer, Gabriella Pitcher, Joshua Hancock, Isabelle Blaber

Within engineering, students are exposed to a high stress academic training environment that can impact their mental health. Further, engineering students who experience symptoms of mental health distress are significantly less likely to seek professional help when compared to their non-engineering peers. Normalizing prioritization of mental health within a College of Engineering has the potential to improve student mental health and help-seeking behavior. Therefore, this study aimed to transition research to practice through implementing mental health training for undergraduate students across a College of Engineering. A 15-minute evidence-based mental health training session was developed and tailored to the engineering experience through integration of quotes from engineering students. The training covered key topic areas including: 1) the importance of prioritizing mental health in engineering, 2) recognizing the difference between normal stress and mental health distress, 3) talking to someone about their mental health and 4) campus resources to support student mental health. To deliver the training, three Counseling Psychology graduate students visited over 50 courses across the College of Engineering with a goal of reaching all students in all years of study across each engineering major. Pre- and post-survey data was collected to investigate the impact of the training on student knowledge about mental health resources and their attitudes about seeking help. This data will be used to support future implementation of mental health training in an effort to normalize conversations around mental health in the College of Engineering.

564c Effectiveness of Academic Supports for Engineering and Computing Education

Edward Gatzke, Robert Petrulis, Sona Gholezadeh

Engineering and Computing students often suffer from high attrition, often due to high levels of academic rigor for elementary level courses. There are many types of academic supports provided to students to help them succeed academically. In addition to traditional lectures and textbooks, students may attend tutoring sessions, meet with academic support coaches, go to supplemental instruction periods, or make use of a variety of other academic supports. Over 200 students responded to an online survey indicating their frequency of use of academic supports. The students also indicated their personal evaluation of the effectiveness of the various support mechanisms. The results of this survey will be presented, including other factors which directly impact student perseverance in engineering and computing majors. Additionally, over 100 academically successful engineering and computing students with high financial need were asked about key factors leading to their academic success. Results and trends from the responses will be categorized and presented. .

564d A Novel Project-Based Course to Promote Chemical Engineering Students' Connecting Agency and Enhance Retention

Ruben D. Lopez-Parra, Madalyn Wilson-Fetrow, Gulnara Kussainova, Konara Mudiyanselag Kanchana Sanjeevani Menike Bandara, Veronica Berry, Sang Han, Vanessa Svihla

The New Directions for Chemical Engineering (2022) highlighted the need for better strategies to foster the recruitment and retention of students from systematically marginalized groups. Retention is impacted by students' persistence, self-efficacy, and engineering identity (Geisinger & Raman, 2013). These individual factors may be negatively affected by students' experiences in foundational courses such as mathematics and physics, where the content is typically taught with minimum reference to engineering and relevant examples (Harris et al., 2015). This disconnection between the course content, the engineering practice, and students' experiences affect students' beliefs and may persuade them to leave engineering. Commonplace remedies require significant course and program transformations. In contrast, we proposed an alternative solution, a new project-based, 1-credit, mandatory course for our first-year chemical engineering students to support their connecting agency. We defined connecting agency as making decisions that connect and situate new or abstract concepts. Namely, students engaged in two projects to help them agentively connect foundational mathematics and physics courses with their prior daily and cultural experiences and the professional engineers' work.

In the first project, students were asked to propose a decarbonized power system for their chosen community in New Mexico. Students chose a familiar community, identified the community energy needs and available resources, and learned about renewable energy through experiments and models. Then, students communicated their proposed power system by presenting it to the community or writing a persuasive letter to a leader. The second project asked students to design a learning activity using sociotechnical problems to contextualize mathematics or physics content that students perceived as unclear and not of value in engineering. Students selected a mathematics or physics concept, explored why it was difficult, investigated its importance, and designed a sociotechnical problem to help others learn the concept. Then, they got feedback from advanced chemical engineering students and shared their final learning activity at an exhibition attended by mathematics and physics instructors.

To assess students' connecting agency, we conducted qualitative analysis of students' artifacts and interviews, and performed basic statistical analysis of a pre-survey and end-of-project surveys. Unsurprisingly, students indicated that they saw few connections between the foundational math and physics courses and their interests and prior everyday experiences. Only 26% reported having opportunities to connect these foundational courses and engineering, even though 75% considered these courses helpful in framing and solving engineering problems. In the decarbonized power system project, students made connections to their roles as civic-minded engineers, their futures as homeowners who could install solar, their preference for problem-solving, and to their interests in sustainability. A few students also reported that what they learned could be useful in their jobs, such as at a science museum. We will expand and contrast these results with students' connecting agency in the second project in the presentation. Helping students develop their connecting agency can support their persistence, identity, and self-efficacy especially through early foundation courses that they perceive as disconnected from engineering. We will share tools and implications for other faculty, including examples of the curricula, surveys, and ways faculty can support students to develop connecting agency.

Geisinger, B. N., & Raman, D. R. (2013). Why they leave: Understanding student attrition from engineering majors. *International Journal of Engineering Education*, *29*(4), 914–925.

Harris, D., Black, L., Hernandez-Martinez, P., Pepin, B., & Williams, J. (2015). Mathematics and its value for engineering students: What are the implications for teaching? *International Journal of Mathematical Education in Science and Technology*, *46*(3), 321–336. https://doi.org/10.1080/0020739X.2014.979893

National Academies of Sciences, Engineering, and Medicine. (2022). *New directions for chemical engineering*. National Academies Press.

564e Broadening Participation: Building Pathways to Success for Transfer Students in Chemical Engineering

Sindia M. Rivera-Jimenez

This presentation highlights the critical role that transfer students from community colleges can play in broadening the participation of underrepresented minorities in science and engineering fields, particularly in chemical engineering. Despite representing nearly half of all U.S. undergraduates, transfer students face unique obstacles and lower transfer rates, disproportionately affecting African American and Latinx students. Therefore, policies and support systems must be established to improve retention rates for transfer students and close the transfer gap.

Student support and advising are crucial for academic success and the overall college experience. Adequate support and advising can assist students in navigating the academic and personal challenges that arise during their time in college, leading to higher retention rates and improved academic performance. This support becomes even more critical for transfer students as they adjust to the new campus culture, understand transfer credit policies, and identify available resources. With appropriate support and guidance, transfer students may find it easier to navigate the complex process of transferring to a four-year institution.

The chemical engineering education community can help support transfer students by providing academic and financial assistance, mentorship, tutoring opportunities, creating transfer pathways, and fostering a sense of community among students. By providing transfer students with the necessary tools and resources to make informed decisions about their career goals, advising can help them prepare for successful post-graduation outcomes. In turn, this support can help transfer students complete their degrees and contribute to developing a globally competitive science and engineering workforce.

In conclusion, this presentation aims to provide a snapshot of the transfer process of community college students to four-year institutions as a vital pathway to broaden the participation of underrepresented minorities in science and engineering fields, particularly in chemical engineering. It intends to spark new discussions and dialogue in the chemical engineering education community about adequate support and advising to overcome their unique challenges and achieve academic and career success.

564f Experiential Learning of Core Chemical Engineering Topics Using Senior Peer Mentoring

Mariajose Castellanos, Neha Raikar

There has been a considerable shift from traditional teaching methods in recent times, and novel practices are being implemented that deviate from the typical in-class instruction. Internships offer a great experiential learning opportunity to integrate classroom learning to real-world situations by helping connect the subject matter to practical applications. Unfortunately, not all students are able to secure internships related to their study area or undergraduate research experiences. In this work, we have created a novel interaction between the final-year students and the juniors by connecting two courses, Capstone Process Design and Process Control and Safety.

The Process Control and Safety course is offered in the Spring semester and is primarily taken by juniors. For the process control project, students who are juniors are hired as interns working alongside the senior design teams. Interested students submit resumes and undergo interviews with all the design teams modeled based on a typical recruitment process. Both parties submit their preferences based on resume evaluation and interview outcome. Instructors for both courses utilize the information and optimize placement for both parties. Respective mentor teams provide the interns with a unique project prompt to be completed over the course of about three weeks.

The deliverables for the project consist of a five to ten minutes video presentation of the work along with all the supporting documents (codes, slides, auxiliary files). The interns are graded on a rubric by the mentoring design team and the two instructors. The categories for grading are: Content, Presentation, Effort, Interaction & Professionalism, and Cooperation.

Some of the learning objectives that the project targets are: 1) Develop and solve fundamental or empirical models for dynamical systems, 2) Understand the fundamentals of control strategies and their implementation, and Use computer-based tools for dynamic model simulation and implementation and comparison of control schemes.

Preliminary feedback on this effort has been very favorable. From the comments, we have seen that the juniors appreciate getting a glimpse of how the material they learn now will be applied in the design. They also enjoy meeting the senior class and getting to know the different projects. In the future, we intend to conduct detailed quantitative and qualitative evaluations of the effectiveness of this effort through surveys and focus group studies. The authors believe this unique effort strengthens the community between senior and junior level classes and benefits overall experience.

564g Alumni Engagement in the Constext of a Capstone Design Course

Joaquin Rodriguez Alonso

Alumni engagement is a most desired target for universities. Colleges generally establish Alumni Associations to foster close ties. Often these associations are structured at the level of Schools or Departments to facilitate a more specific relationship. It is recognized that they provide prestige, donations, advising, and mentoring of high impact. Special awards, fundraising campaigns, membership in advisory boards, and networking are among the many activities promoted by colleges to keep in contact with alumni, and benefit from their participation. However, there are limited approaches to engage alumni with curricular courses. Some experiences include invited talks and participation as judges in capstone design projects. We have developed a new initiative in the context of the capstone design courses in chemical engineering. Senior students are paired with alumni by the instructor or by student self-selection. Students request a first meeting with partner alumni after some biographic research and the introductions by the instructor, explaining the scope of the project. Students prepare some questions for this meeting to elaborate a bio-sketch (first deliverable). Partners are open then to follow up with meetings at their convenience. One theme for conversations is the experience at college, both academically and personally. The content of the capstone course is a common anchor topic as alumni usually remember the project they ran. The goal is to produce a short assessment (2-3 pages) contrasting both experiences and offering some advice to improve the educational performance at the department (second deliverable). Partners are invited by the instructor to discuss freely on career development, work-life balance, professional skills, etc. Students elaborate on the learning experience producing a short assessment with reflections and major take away (third deliverable). In addition, students are invited to participate in focus-group meetings to briefly share the experience with some classmates and provide useful information to evaluate and improve the project. This paper describes the experience of the last two years with this project, involving 98 senior students and over 80 alumni. Alumni graduation date range extends back to 1978, with increasing participating of more recent graduates. Quantitative assessment provided in the paper reports that this project is one of the top engagements of alumni and very often their only connection with the university. Senior students confirm the appreciation for the project with impacts including choices in career orientation and job offers.

564h Cultivating Pathways to Graduate School: Investigating the Impact of Peer Mentoring on Hispanic Engineering Students

Paola Baldaguez Medina, Federico Cifuentes-Urtubey

Mentorship is critical for fostering academic and leadership development among underrepresented groups in STEM. Research shows that strong peer relationships, effective communication skills, and favorable academic and career development factors are vital in encouraging students from underrepresented backgrounds to pursue advanced degrees and careers in engineering. To further support the pursuit of advanced degrees by underrepresented populations, we propose that graduate students can play a crucial role in mentoring undergraduate students, thereby providing guidance towards graduate-level degrees and advanced career opportunities. We seek to enhance Hispanic representation in graduate school through a hybrid mentorship program that analyzes the factors enabling successful mentoring. Our program focuses on cultivating belonging within the field of engineering by bringing together Hispanic graduate students and undergraduates. Participants had the flexibility to meet virtually or in-person. In the first iteration, 12 graduate students mentored 12 undergraduate students from various departments in the Grainger College of Engineering at the University of Illinois. The program enabled 50% of graduate student participants to gain their first experience as mentors. By the end of the first iteration, 91% of graduate mentors expressed interest in acquiring more skills to better support their mentees. Although none of the undergraduates in the first cohort were eligible to apply to graduate school, two of the three senior undergraduates in an ongoing second cohort have applied to graduate programs or fellowships. Our program underscores the importance of representation, guidance, and monitoring in successfully boosting students' interest in graduate school participation. By incorporating mentorship into chemical engineering classrooms, students can receive the guidance and support they need to succeed academically and professionally. Such efforts can help increase diversity and representation in the field, especially for underrepresented groups in STEM.

564i Cocreation of of Inclusive Leadership Training with Undergraduate Engineering Students

Ingrid Paredes

Undergraduate teaching assistants play a key role in student success and mentorship of their cornerstone design project. As the first points of reference for students, they assist with content delivery, guide students through hands-on labs and projects, and deliver regular feedback on assignments. Effective teaching assistants are leaders and peer mentors and their training as educators is essential to our students' success. Each semester, the undergraduate teaching assistant community for introductory engineering organizes peer-facilitated training on course content, technical skills, and best teaching practices. The training is grounded in global inclusion, diversity, belonging, equity, and access (GIDBEA) to foster a sense of belonging among the community of teaching assistants, students, and faculty. To this effort, we are piloting a series of workshops on inclusive leadership to be delivered ahead of each semester.

We seek to build our teaching assistants' sense of agency in the classroom by cultivating a positive self-concept, developing their understanding of sociopolitical environments, and providing resources for action. Co-created with faculty, teaching assistants, and DEI experts at the institution, the workshop series provides teaching assistants with the ability to recognize and confront bias among individuals and within teams, helps them develop an understanding of power, privilege, and oppression, and equips them with the tools to employ their knowledge professionally. The workshops feature individual reflection activities and small group discussions, culminating in a community-wide discussion on lessons learned and actionable items to build an inclusive community within engineering programs.

To understand the value that this training provides the teaching assistants, a survey was conducted of participants before and after participation in the workshops. The survey aims to capture the practicality of the training and the teaching assistants' assessment of the climate within the program. In this complete experience-based practice paper, findings from the second year of piloting our workshops are described. In this second iteration of training, new teaching assistants participated in our foundational training in GIDBEA, and returning TAs built on their introductory knowledge to learn about social justice and principles of inclusive leadership. The data shows that all teaching assistants overall found the workshop content and activities were relevant to them as peer educators. Several teaching assistants shared inclusive leadership strategies that they planned to implement in the coming semester. The goal of this study is to inform plans for implementing solutions into training that address deficiencies identified through the survey and provide a set of inclusion best practices and learning objectives for inclusivity training for undergraduate teaching assistants.

348b Using an Entrepreneurial Mindset to Construct a Winning Chem-E-Car

Kevin Fruth

Each year, AIChE conducts the Chem-E-Car Competition that tasks teams of college students with designing and building a shoebox-sized car powered by a chemical energy source that will safely travel a specific given distance and then stop utilizing a chemical reaction. In 2021, the Univ. of Toledo (UToledo) Chem-E-Car Team exhibited the entrepreneurial mindset through the application of curiosity, connections, and creating value in the construction of the zinc-nickel battery for their Chem-E-Car. They earned first place in the Annual Student Conference Chem-E-Car Competition (earning the H. Scott Fogler trophy) by stopping just 10 cm from the target distance of 21.05 m.

348c Entrepreneurially Minded Learning in the Chemical Engineering Unit Operations Laboratory

Erick S. Vasquez

Over the past few years, chemical engineering students in the unit operations laboratory have been encouraged to adopt an entrepreneurial mindset (EM) to solve non-traditional unit operations experiments. Specifically, the KEEN framework of the 3 C's, curiosity, connections, and creating value, has been implemented to work on final open-ended projects with sponsors outside the University. It has been demonstrated that students' self-directed learning correlates with EM when working in the laboratory. This talk briefly discusses final project assignments, deliverables, assessment methods, stakeholders, benefits, opportunities, and limitations.

348f Embracing the Entrepreneurial Mindset: Transforming Academic Research into Tech Start-Ups

Chelsea Monty-Bromer

Faculty members are exploring start-ups to grow and expand their research interests. In this talk, Chelsea Monty-Bromer will discuss her journey from academia to entrepreneurship with her start-ups, SweatID and Corrolytics, demonstrating how adopting an entrepreneurial mindset can lead to success in securing funding, creating job opportunities for students, and contributing to the growth of the tech industry in Northeast Ohio.

259a Prioritizing Learning Objectives for Chemical Engineering Laboratory Courses: An Industrial Perspective

Sarah Wilson, Samira Azarin, Christopher Barr, Joanne Beckwith, Janie Brennan, Tracy Carter, Amy Karlsson

Since March 2020, many chemical engineering laboratory instructors have been forced to examine what it means to be a chemical engineering lab or more specifically a Unit Operations laboratory. Historically, chemical engineering laboratory courses are taught in the third and/or fourth year and expose students to types of equipment they may see in industrial positions while also giving them experience in interpersonal skills such as teamwork and communication. However, there is little guidance on what should be assessed in the chemical engineering laboratories. Previous work by Feisel and Rosa has suggested thirteen general learning outcomes that can be covered in engineering laboratory courses. However, it is nearly impossible to assess students' understanding for all thirteen of these learning outcomes. Additionally, these learning outcomes were neither specific to chemical engineering labs nor to the needs of the stakeholders of chemical engineering labs.

To work towards alignment of laboratory courses with industrial needs, a better understanding of the learning outcomes currently addressed in laboratory courses and the gaps in the current curriculum need to be identified. A survey was designed to gain an understanding of the outcomes important to the various lab stakeholders (students, faculty and industry engineers) and the current success of chemical engineering laboratory curricula in addressing those outcomes. Results will provide insight into the prioritization of laboratory learning outcomes and allow for the redesign of laboratory courses to better align with the skills and attributes desired by employers.

This presentation will describe responses received by the engineers in industry. The 71 responses include demographic and background information, which can be used to situate the survey responses in context of the respondents' experiences. Additionally, respondents were asked to rank the five most important learning outcomes for chemical engineering laboratory courses for a new hire in chemical engineering, identify which outcomes new chemical engineers are weakest in, and which outcomes they feel the chemical engineering curriculum should do a better job covering. Finally, open-ended questions were included to identify additional important learning outcomes and provide comments.The responses provide important context for the engineering laboratory since a majority of chemical engineering students graduate and continue into a career external to academia.

Reference: Feisel, L. D. and Rosa, A. J., The Role of the Laboratory in Undergraduate Engineering Education, Journal of Engineering Education, 94(1), 2005, 121-130.

259b Incorporating Skills Experiences in Unit Operations

Troy Vogel, Victoria Goodrich

Lab for Chemical Engineering students at Notre Dame consists of a junior lab (Lab 1), and a senior lab, taken in the spring and following fall semesters, respectively. All students take the same junior lab and there is a curricular option for senior lab where students choose between a more traditional Chemical Engineering lab (Lab 2) and a Biomolecular Engineering lab (Biolab). In spring 2023, junior lab was changed to incorporate 6 skill development experiences. Before changes, junior lab consisted of experiments, 3 weeks in duration, where students in groups of 3 would rotate through 4 of the 4-7 experiments running that semester depending on the number of students taking the course. A prelab and safety quiz is required at the start of week 1. A preliminary report with specific deliverables is due week 2 in order to catch computational errors or gaps in understanding. Some experiments have a deliverable due at the beginning of week 3. A final report is due during 'week 4' or week 1 of the next experiment rotation. Students then have an oral defense of the work with the instructor, usually within about one week. The final rotation deliverable is a shorter, extended abstract and public presentation with Q&A from all the lab faculty members, replacing the final report and defense.

The change to Lab 1 involved removing the first of the 4 cycles, replacing the cycle with 6 skill development experiences, 2 per week. Reasons for including these skills experiences as well as selecting the specific skills will be presented. The skill development experiences were: group based (group formation and group contracts), lab safety audit, technical writing instruction, calibration and non-linear modeling with error analysis, creating a piping network, and reading a P&ID. A short summary (with pictures!) of each skill development activity will be shown.

Shortly after the 3 week skill development cycle, a 5 point Likert survey was given asking students about desirability of each skill development activity and self-perceived usefulness of each skill with regard to their lab experience (students currently taking Lab 1 with Lab 2/Biolab taken the following fall semester) and future professional career. After the conclusion of Lab 1, the same survey will be administered again, only changing the wording to reflect realized usefulness of each skill in Lab 1. Comparison of these results will be provided.

Early results are not surprising. Overall, students enjoy the activities that have a concrete goal of building something the most. The student perceptions of usefulness of each skill in their two lab courses and professional career loosely follow the instructors' intent.

259c Teamwork Skills Improved By in-Class Team Building Games in Chemical Engineering Lab Courses

Joaquin Rodriguez Alonso, Michael McMahon

Teamwork skills rank at the top of the list for employer's demands of professional abilities for new hirings and further in the career. Chemical engineering curricula at many universities provide for extensive teamwork skills development by including team projects in several courses, most distinctively at capstone design courses. This training provides students with some confidence and references for approaching their work in industry and academia. However, these experiences are mainly based on requesting students to act together to complete some complex tasks (assignment, project, presentation, etc.). ABET Criterion 3 Outcome 5 requests from accredited programs to document that students develop "an ability to function effectively on a team whose members together provide leadership, create a collaborative and inclusive environment, establish goals, plan tasks, and meet objectives." Some efforts have been reported on providing students with instructional content on team work, performance, and leadership to meet these requirements. However, there is limited evidence on the efficacy of these approaches. In our experience, students usually limit themselves to the "divide and conquer" strategy of distributing tasks for results to be assembled generally close to the due dates. Valuable tools like team contract, planning spreadsheets, progress monitoring, log books, leadership strategies, and synergistic communications are frequently missed. Some of these concepts, even if introduced in first-year programs, do not carry lasting impacts throughout the curricula. We have introduced in-class team games, specifically addressing important teamwork skills, as engaging activities to reinforce key components of effective team performance. This paper reports the experience of one of these team games with laboratory courses for sophomores (52), juniors (61), and seniors (71). Quantitative results are analyzed by scorecards and reflections by students to assess the impact of this active learning strategy. They also provided a valuable source for developing new team games addressing the weaknesses and problems detected by students and instructors.

259d Teaching Students How to Develop Hypotheses for Experimentation in the Unit Operations Laboratory

Stephanie Loveland, Kristin Stoner

The Unit Operations Laboratory is a critical course in the education of chemical engineering students. Very often, this course includes experiments that are prescribed, in which students follow standard operating procedures, make observations, and answer specific questions [1]. The results are reported using various methods, which can include written reports and oral presentations. The students' understanding of the process is demonstrated in their presentation of results and ability to analyze data according to the theories they have presented and tested in the laboratory.

To encourage better understanding of the processes studied, students can approach the experiments with a set of more open-ended questions to answer. This will require students to design their own set of experimental tests to be conducted. This experimental work must follow from a logical set of hypotheses, based on the theories to be tested [2] [3] [4]. Formulating a good hypothesis will inform their experimental design and lead students through their experimental testing in a logical fashion.

In our experience, students have struggled to formulate reasonable hypotheses that are not superficial or "obvious" from basic observations and knowledge. To address this weakness, we have developed a set of activities and rubric criteria to give students guidance on how to formulate a good hypothesis for testing in the Unit Operations Laboratory.

In this paper, we will present one of those activities as well as the rubric language developed to support student formulation of strong hypotheses statements for experimental testing. Examples of typical student hypotheses (before and after the workshop) for various unit operations experiments will be shared and discussed. Survey results from students who participated in the activity will also be shared to show the perceived effectiveness of the activity.

References

[1] A. E. Ghadi and R. Mammucari, 'An acve laboratory learning experience for chemical engineering students facilitated by hypothesis tesng,' in Research in Engineering Education Symposium & Australasian Association for Engineering Education Conference, Perth, 2021.
[2] B. R. Young, H. W. Yarranton, C. T. Bellehumeur and W. Y. Svrcek, 'An Experimental Design Approach to Chemical Engineering Unit Operaons Laboratories,' Education for Chemical Engineers, vol. 1, no. 1, pp. 16-22, 2006.
[3] P. C. Price, R. Jhangiana, I.-C. A. Chiang, D. C. Leighton and C. Cutler, Research Methods in Psychology, Creave Commons Internaonal, 2017.
[4] S. Mcleod, 'Research Hypothesis: Definions, Types & Examples,' 6 March 2023. [Online]. Available:
htps://simplypsychology.org/what-is-a-hypotheses.html. [Accessed 30 March 2023].

259e A Question of Scale: Equipment Size in Teaching Labs

LiLu Funkenbusch, Fernando Mérida Figueróa

Our primary goal when teaching a laboratory course is to provide hands-on applications and experiences with the concepts that students learn in the classroom. The first questions we ask ourselves are which concepts and what experiments. These choices are largely based on the specific industries that hire our students or research specialties of the faculty who teach or have taught the course.

The next question is then the scale of the experiment. The best educational experience is often said to match the scale of processes they will use in industry. This is usually prohibitive, in terms of facilities, lab space, maintenance/upkeep costs, safety, and run time. Out of necessity, we therefore shrink our equipment to pilot or smaller scale operations.

At the University of Florida (UF), we have a wide range of experiments in desktop/miniaturized, benchtop, and pilot-scale. In this talk, we will discuss the experiments we have at UF, student feedback for each scale of experiment, and the advantages/disadvantages that the scale of each experiment offers.

259f Design of a Process Safety Focused Tank Purging Process for a Chemical Engineering Unit Operations Course

Sarah Wilson, Tracy Carter, Nick Cprek

Within a chemical engineering unit operations course, students are often asked to apply their knowledge of process safety to understanding the safety of the experiments in the laboratory. While this is valuable in allowing students to see the application of process safety across different pieces of process equipment, most experiments in the lab are not explicitly designed around process safety concepts. Therefore, this work aimed to design an experiment for a unit operations laboratory focused on the process of purging a tank for flammable fuel storage and transfer. In industry, nitrogen can be used to purge systems to ensure oxygen levels are maintained below the limiting oxygen concentration to prevent fires and explosions. The designed experiment features two tanks that can be used to demonstrate the process of purging a storage tank and transferring a flammable liquid between two storage tanks. Students can determine the cycling requirements for tank purging as a function of target oxygen concentration, as well as the low and high pressure for the cycle. Additionally, they can determine the volume of nitrogen required to reduce oxygen concentrations when purging the system under atmospheric conditions. Through this experiment, students are exposed to important, industrially relevant process safety concepts and can compare experimental data back to theoretical models for pressure cycling and sweep-through purging. This presentation will focus on the design of the equipment, as well as present experimental data fit with theoretical models. Through design of experiments on process safety, students can increase their knowledge of process safety concepts and theory in a unit operations course.

259g Beer Brewing As a Practical Application of Chemical Engineering Fundamentals in Lab and Design

Anastasiya Islamova, Iram Rahman, Benjamin Davis, Jennifer Weiser

The craft beer industry is growing exponentially due to increased consumer interest in homebrewing and new styles of beers; craft beer sales volume grew 8% in 2021 and retail sales of craft beer account for almost 27% of the $100 billion U.S. beer market. [1] Learning about the brewing process as an undergraduate chemical engineering student provides opportunities to better understand bioreactor design, process control, and reaction engineering in a more relatable and interactive way.

Brewing involves several key steps (including malting, milling, mashing, and hop leaching) with the main process step being fermentation, where chemical conversion of sugars into carbon dioxide and ethanol occurs through the growth of yeast in a bioreactor. All of these steps require close monitoring and control of key parameters including temperature, pressure, and pH in order to produce a high-quality product. By learning about these processes as undergraduate student and having a hands-on lab experience, students gain valuable knowledge on the effects that key parameters can have on the final products and how they relate to industrial brewing. This knowledge translates to other bioreactions in the chemical engineering field including the production of pharmaceuticals, biopolymers, perfumes, and other bioproducts. [2]

We created a beer brewing experiment in the senior chemical engineering laboratory course which focuses on the fermentation aspect of brewing; students use dry malt extract (DME) to create a wort that is boiled, cooled, and fermented. The experiment is aimed at investigating the effect of different parameter changes and process controls in several small-scale bioreactors of various sizes and capabilities on the final alcohol by volume (ABV) of a beer. The experiment is inquiry-based: students are encouraged to experiment with different ingredients, concentrations, or brewing techniques to allow for creativity in the number of factors that affect the quality of the final product. Important QC/QA criteria like specific gravity and temperature are measured in real time using a tilt hydrometer and students also learn to characterize their product in other ways including final pH and/or yeast cell count/turbidity. Students learn to model the kinetics of the fermentation reaction via the Monod kinetic model and aim to model and validate with data things like biomass accumulation, ethanol production rate, and substrate uptake. They also use a refractometer to do live cell counts and can use their results to compare to their kinetic model.

Additionally, the senior capstone project for the same cohort of students asked groups to do a conceptual design of a manufacturing facility for production of either 1,000 or 15,000 barrels of beer per year and some other co-product (either biochar or PHA bioplastic). Students were asked to produce both the beer and the co-product with the overall sustainability of the process in mind, considering the cultural, social equity, technical, economic, and environmental impacts of the new process over its entire lifecycle. We aim to present the results of student work from the most recent instantiation of the lab and design courses and to demonstrate the broad appeal of brewing as an application which is relevant to students.

[1] Craft Begins Recovery Amid Supply Chain Challenges. *The New Brewer* 44–48 (2022).

[2] Benz, G. T. Bioreactor Design for Chemical Engineers. *Chemical Engineering Progress* (2011).

259h Experiments in Process Dynamics and Control at the University of Kansas

Felipe Anaya, Mark B. Shiflett

A novel, state-of-the-art Process Dynamics and Controls experiment was recently commissioned, constructed, and implemented for the Unit Operations Laboratory (UOL) at the University of Kansas (KU). The addition of this experiment is part of a concerted effort to update, improve, and modernize the UOL facilities with the chief goal of improving student learning by providing laboratory equipment, instrumentation, and experiments relevant to the practice of the profession today. The UOL course at the KU is designed as a series of two courses offered in the Fall and Spring semester of the Senior year. In the Fall semester students carry out experiments on kinetics, thermodynamics, and fluids and in the Spring, they focus on distillation, absorption, and process Dynamics and controls. In this format, students can focus on one experiment for a period of 5 weeks in which they visit the lab twice, make a presentation about their experimental plan, and prepare a written technical report. Students have provided positive feedback about the course structure and the quality of the data collected. The changes in both the course structure and laboratory facilities have yielded positive results on student learning and attitudes toward the laboratory course.

In this presentation we will discuss our learnings from offering the new process dynamics and controls experiments to students in the laboratory course. First, students visit the lab during the "pre-lab tour" and learn how to operate the equipment and collect preliminary data. They use this data to design a control strategy, build a model for the system, and solve the model using Simulink, python, or similar program. Next, they prepare a presentation about the experiment principles and theory and discuss their experimental plan. This include the goals for the experiment and the list of trials to be completed. Finally, they return to the lab and carry out their data collection. A preliminary report submission is due a week after the lab day and the final written report is due two weeks after the lab day. In the last few offerings of the course students have proposed and study different schematics for liquid level control in and explored experiments in temperature control. Student feedback will be presented and discussed in light of the course outcomes and experiment specific outcomes.

233a Guiding and Motivating Students' Practice of Fluid Dynamics Concepts Using a Token Economy System.

Sanha Kim, Roseanne Ford

As a discipline, chemical engineering often presents students with concepts that are novel or highly specialized applications of foundational chemistry and mathematics. To support students as they build their conceptual understanding, class instruction is paired with opportunities, such as homework assignments, for students to practice course material. However, the fast pace of core curriculum classes that is seemingly necessary to cover all required concepts can limit the amount of "low stake" homeworks that cover a specific topic. Without sufficient practice and subsequent instructor feedback, students will likely find recalling information and effectively applying concepts difficult to achieve during "high stake" quizzes or exams. Although the instructor can still direct students to resources that offer more topic examples, there is rarely a formal class system in place that helps students receive timely feedback on such additional practice or helps motivate students to utilize such resources.

Here we explore how a token economy can guide and motivate students to revisit concepts throughout an undergraduate fluid dynamics course. Under this system, students are able to acquire tokens as they fulfill previously expected class participation and pre-lecture preparation. Once tokens are acquired, students are able to exchange tokens for resubmissions of prior homework or quizzes with commentary about how the student has resolved their errors. If the resubmission is accepted, the student recovers a portion of their missed points for that assignment's final grade. The effects of the token economy on the course experience will be assessed through student survey responses with the focus around how the token economy influences 1) perceived class workload, 2) student motivation for revisiting assignments, and 3) perceived utility of homework for exam preparation. We hope that the implications of this study will highlight the token economy as an instructional tool that can easily be integrated into a core chemical engineering course and promote students' conceptual practice outside the classroom.

233b Implementing a Collaborative Online International Learning (COIL) Project in a Chemical Engineering Fluid Flow Course

Erick S. Vasquez, Erick Ramos

Video conferencing tools allow collaborative opportunities among students and faculty in Chemical Engineering worldwide. A well-established approach to engaging students from different institutions or countries in a remote setting is known as collaborative online international learning (COIL). An attractive feature of COIL is the low-cost and travel-free opportunity for students to engage with a different culture directly. Different levels of interactions are possible with this type of pedagogy and can vary from a few weeks to an entire semester. The unique interactions can also provide cross-cultural learning experiences under the right conditions.

This work discusses several aspects of a COIL project involving student participants from the University of Dayton (Dayton, OH) and Universidad Centroamericana, UCA (El Salvador). Students were placed in teams of four and had approximately three weeks to work on the theoretical design of a viscometer with certain limitations. Before working on the theoretical viscometer design, students recorded short introductory videos (~ 2 – 3 mins) including name, location, birthplace, hobbies, the reason(s) for choosing chemical engineering as a major, or any other information that the student wished to disclose and the expectations for this COIL activity. Students, for example, were also tasked to describe their favorite place on their campuses by sending pictures to the online forum. Students in both countries had complete access to the University of Dayton Learning Management System (UD LMS).

Discussion forums were also available online for each participating team. First, to initiate the conversation, students brainstorm ideas individually and then share them with their teams. Flexibility was a vital aspect of this COIL activity. Differences in student enrollment, time zone, languages, and semester duration were determined initially to establish the project's timeline. Next, most students decided on the communication level via video conference or texting. Next, the instructors requested meeting notes or screenshots to confirm that the teams interacted throughout the project, and both instructors provided continuous feedback to all the teams. Lastly, students had to record a technical video presentation about the design of the viscometer following a Pechakucha video-style presentation. This presentation format has 20 slides with 20 seconds for each slide (automatically timed), totaling 6 minutes and 40 seconds. Slides should be primarily pictures, figures, tables, or drawings, which could include equations.

Most of the COIL teams performing the viscometer design had positive interactions and enjoyed the activity as they learned about students studying chemical engineering in another country. However, miscommunication among team members occurred in a few cases, leading to frustration. Lastly, we discuss potential aspects to improve for future COIL implementations.

233c *Applied Statistics Learning Modules for a Chemical Engineering Transport Lab Course*

Joanne Beckwith

As part of a course redesign, explicit instruction and activities relating to statistics and design of experiments (DOE) were incorporated into a third-year undergraduate chemical engineering laboratory course. Development of the modules was motivated by exit interviews with senior students, course evaluations, and faculty surveys which identified a need for students to apply statistical tools in a meaningful way with "real-world" data. This work serves as a foundation for a realignment of the two-course chemical engineering lab series at our institution.

Taking an active learning approach, statistical analysis and DOE were taught using student data acquired in multiple transport process experiments with increasing open-endedness throughout the semester. Students used both a factorial design, and a Latin Square block design to simultaneously change multiple variables across multiple trials, demonstrating a real-world example of applied statistics. The modules are guided by the following learning objectives: By the end of this course, students will be able to (a) fit a statistical model to a given or collected data set (b) describe what DOE is and how it is used in a lab setting (c) describe DOE tools (Block Designs, Factorial, etc.) and identify under what circumstances we would use such tools. Lab reports were coded for statistical outcomes and a pre-test / post-test assessment assessed the efficacy of the learning module. Additionally, registrar information on previous statistics courses taken, and demographic information were obtained to situate the impact of the learning outcomes. After the successful implementation of this module, this work will be expanded to include the student application of DOE in the fourth-year unit operations lab.

233d Pipe-Flow Challenge: An Open-Ended, Experiential Engineering Module Based on Fluid Dynamics

Jacob Borden

As engineering educators, we are always wanting for lab modules that are some combination of interesting, challenging and feasible. As chemical engineering educators specifically, we also need more laboratory exercises that require students to build.

The pipe-flow challenge is a fluids-based, open-ended laboratory where students practice the application of science and math for building a fluid network that achieves a design objective within certain constraints. As important, the components for this lab are easily and cheaply acquired from any local hardware store.

The pipe-flow challenge requires components including valves, quick connects, and straight tubing typically found in every day domestic plumbing. Our lab makes use of the PEX line of tubing, fittings and gauges, and a Barracuda ½ horsepower model pump. When connected with a regular sink faucet, this system is ready to evaluate pressure / flow relationships in order to understand fluid dynamic calculations.

Students apply science by testing and measuring system pressures relative to flow rates in order to characterize, e.g., pump curves, pipe roughness and friction factors, and K-values for fittings and valves. Students apply math using best-fit regression between measured and calculated pressure drops based on appropriate fluid dynamic friction equations. Finally, students are required to build a fluid network to deliver a certain number of gallons of water within pre-defined constraints. Constraints may include a designated pump running time, a minimum or maximum number and type of fittings and pipe / diameter combinations, and distance and elevation change requirements.

The pipe-flow challenge has many potential extensions. For instance, packed beds can be made with glass beads of varying size and shape, with pressure/flow data to estimate best fit parameters for the Ergun equation. With a plug-in power meter to measure pump voltage and current, parameters including pump power and efficiency can be included in either the objectives or the constraints. The pipe-flow challenge also spans multiple disciplines and is practical for both civil and mechanical engineering students.

The pipe-flow challenge was first incorporated into the Unit Operations lab at Trine University in 2021. This talk will present on the materials, methods, and design objectives developed over the past three iterations of implementation. By attending, audience members will appreciate the materials and methods as well as learning objectives and outcomes associated with applying the pipe flow challenge on their own campuses.

233e Optimizing Oil Pipeline Flushing during Product Changeovers: A Bench-Scale Pilot Plant Study

Barnabas Gao, Swapana Jerpoth, David Theuma, Sean Curtis, Michael Fracchiolla, Kirti Yenkie, Robert Hesketh, C. Stewart Slater, Mariano J. Savelski

Flushing oil pipelines during product changeovers is a critical process in lube oil processing facilities to ensure the final product's quality and purity.his flushing operation results in large volumes of comingled products that are not packaged and sold at the high value of the pure products. So not only is there a financial loss, but the comingled oil product is not packaged which lowers the product yield. To reduce both the financial and material loss an innovative approach to optimizing this process is required.

We have conducted a limited number of studies at the industrial plant, but need more data to optimize this process. These industrial studies were found to be both time-consuming and labor-intensive for an operating lube facility. It was found that these studies required substantial changes to the operating procedures in the plants and also resulted in excess downtime at the plant. To address these issues, a laboratory-scale pilot plant was designed and built to study and optimize the flushing process.

A comprehensive review of lube oil properties, pipeline operations and flushing processes in the lube oil industry will be studied with the pilot plant. Using the pilot plant, we are able to study the mixing of a residual oil with a flushing oil in a pipeline system. The pilot plant has been designed at a geometry ratio of 1/5 of the industrial plant. The flowrates in the pilot plant are based on matching the Reynold's numbers of the industrial plant. One advantage of the pilot plant is that we are able to vary the flowrate of the oil in the system using a positive displacement pump with a variable speed drive. This allows us to examine flows of identical Reynolds numbers to the industrial plant and examine an extended operating range of these flowrates (Scoffoni et al., 2001; Wang et al., 2011).

With this pilot plant we can not only obtain fluid samples and measure kinematic viscosity, similar to the industrial plant, but we also have added an online measurement of the viscosity. We are also able to examine various pipeline system configurations with the pilot plant. This includes alternative filters, fittings and piping. Also, in this pilot plant we have studied pipeline coatings to determine if they will reduce the amount of residual oil that needs to be flushed. The research findings from this pilot plant are providing valuable data that are being used to formulate a model of this process.

The bench-scale pilot plant provides a valuable tool for studying the flushing of oil pipelines during product changeovers in lube oil processing facilities. The novel approach will aid in the validation of models and the optimization of the flushing process, resulting in increased efficiency and product quality. The ability of the pilot plant to simulate industrial flush operations on a smaller scale provides a cost-effective and efficient solution for studying and optimizing the process while reducing the need for excessive operator time and plant visits.

REFERENCES

Scoffoni, J., Lajeunesse, E., Homsy, G.M., 2001. Interface instabilities during displacements of two miscible fluids in a vertical pipe. Phys. Fluids 13, 553–556. https://doi.org/10.1063/1.1343907

Wang, S., Huang, X., Yang, C., 2011. Mixing enhancement for high viscous fluids in a microfluidic chamber. Lab. Chip 11, 2081–2087. https://doi.org/10.1039/C0LC00695E

233f 3D-Printed Modular Membrane Frames for Student Projects and Class Demonstrations

Daniel Anastasio, Marissa Tousley, Heather Chenette

Various membrane separation technologies are gaining prominence industrially, so chemical engineering educators may be seeking ways to demonstrate membrane separation principles to their students. However, these membrane systems are often expensive, difficult to modify, and may only be able to demonstrate a single separation technique, collectively limiting access to hands-on membrane experiences and the learning objectives the equipment can fulfill. This presentation will showcase a low-cost, desktop-scale plate-and-frame membrane separation system that can be easily modified for demonstrations of most low-pressure membrane processes. The membrane frames were printed using a stereolithography (SLA) 3D printer and designed to be modular and modifiable, with parameters such as dimensions, support structures, and flow paths being easy to change using a CAD software. Flow is driven using a series of small aquarium pumps controlled by an Arduino microcontroller. These modules can be used for capstone laboratory experiments or in-class demonstration of membrane separation concepts. We will show the results of a student laboratory project that examined the impact of voltage on separation efficiency in electrodialysis. We will also show how we used this platform in our project-based membrane elective course developed by the authors. In the course, the modular platform was used to demonstrate the phenomenon of concentration polarization in forward osmosis and used by students to conduct short-term research projects.

233g A Flexible Membrane Experiment to Teach Gas Separations in the Laboratory and in the Classroom

Carlos Landaverde-Alvarado

For the last few years, the department of Chemical Engineering at the University of Texas at Austin has committed efforts and resources to the renovation of its undergraduate teaching laboratory sequence in the Junior and Senior years. These efforts have pursued a problem-based-learning focus and a continuous improvement approach to improve the practical educational experience for every chemical engineer who graduates from the program. As part of this renovation, we have greatly benefited from our collaborations with and the input we receive from industry, and the participation of staff, faculty, and graduate students in the development of the new learning experience offered to our students. To this day, we have added, redesigned, and improved all the experiment stations in the laboratories with aims to enrich the learning experience in the chemical engineering program.

Chemical engineering laboratories are courses in which students solidify their knowledge and put theory into application by working in teams to solve challenging problems. In these laboratories, undergraduate students strengthen their understanding of fundamental chemical engineering concepts introduced in courses such as Transport Processes, Fluid Mechanics, Material Balances, Thermodynamics, and Separations amongst others. Many of these fundamental concepts can be demonstrated effectively when applied to gas separations and membrane systems applications.

Membrane gas separations are widely used in a vast array of applications as they provide advantages over other separation techniques such as small size and portability, long-term cost stability, and modular nature for future expansion. If applied inside the classroom, these industrial process advantages can be translated into experiments that provide a flexible and tailored laboratory learning experience. Thereupon, we have developed a gas membrane separation experiment that offers ample design flexibility for laboratory and classroom learning applications. The membrane system consists of 4 interconnected and easily exchanged small-scale membrane modules that facilitate the study, design and scale-up of membrane separation processes. The system is equipped with multiple gas sources to facilitate gas mixing on the fly, it has interchangeable rotameters and flow meters to scale up or down the process, and we have installed multiple gas analyzers to study different gas separation applications (e.g. O_2/N_2, CO_2/Air, etc.). In addition, the experiment has been assembled in a mobile workbench (i.e. "on-wheels") to facilitate its transport for use in classroom demonstrations and in engagement activities during recruitment events.

In this presentation, we will describe this newly developed gas membrane separation experiment station and its contributions to our continued development approach. We will detail the design and construction of the experiment, its relationship to chemical engineering core courses, its use as a flexible class project in laboratory courses, and its application in demonstrations outside the laboratory. We will also share example projects developed by students in our Senior Laboratory to highlight how the practical elements of this experiment contribute to the overall educational experience in the chemical engineering program.

233h Developing a Project-Based Elective Course on Membrane Separations

Heather C. S. Chenette, Marissa Tousley, Daniel Anastasio

Undergraduate chemical engineering curricula rarely offer courses spotlighting membrane separations (1). Courses that include hands-on project-based learning benefit students yet can create a challenge for faculty as they can require additional preparation to develop, implement, and assess. This presentation describes a project-based elective course on membrane separations geared toward junior- and senior-level students in chemical engineering or related fields, and it's evolution over three iterations. We will share a five-unit framework for introducing membrane topics that resolves challenges associated with interweaving membrane applications with their separation and fabrication principles. We will also discuss how the course has evolved to integrate and align laboratory activities with lecture content, include scaffolded open-ended projects and opportunities for peer-instruction, and leverage the expertise of multiple instructors. Additional information on the novel membrane systems used in this elective course will be provided in another talk, '3D-Printed Modular Membrane Frames for Student Projects & Class Demonstrations'.

1. Vigeant, M. A., & Silverstein, D. L. (2014, June), *Results from the AIChE Education Annual Survey: Chemical Engineering Electives* Paper presented at 2014 ASEE Annual Conference & Exposition, Indianapolis, Indiana. 10.18260/1-2--22983

217a Professor L.S. Fan's Legacy at Ohio State

Umit Ozkan

During his 45-year tenure at Ohio State Chemical Engineering, Professor Fan left his mark at every aspect of the department. As an educator, as a researcher, as a leader and as a mentor, he led the department to new heights and set very high standards for his colleagues and his students to aspire to. After 45 years, he continues to shine as a bright guiding light for all of us. Reflections on his impact on our profession and on the Ohio State Chemical Engineering will be shared.

217b Prof. Liang-Shih Fan's Legacy at Aiche

Darlene Schuster

Prof. Liang-Shih Fan has provided remarkable leadership and mentorship to AIChE and its members. It is our honor to celebrate his outstanding contributions with the 2023 Warren K. Lewis Award, which recognizes distinguished and continuing contributions to chemical engineering education.

217c Prof. L.-S. Fan's Impacts on Industries - in Honor of the 2022 Receipt of the Warren K. Lewis Award

Bing Du

Prof. L.-S. Fan received the 2022 AIChE Warren K. Lewis Award to recognize his distinguished and continuing contributions to chemical engineering education, which is one of the most prestigious awards for education in chemical engineering. Prof. Fan has published many high-quality journal papers, patents and books including the widely adapted textbooks. He also spends extensive efforts for educational outreach activities, i.e., in various academic societies, science and engineering fairs, advisory board/editor of multiple scientific journals, mentoring young engineering faculty, advising over 200 graduate students and post-doctoral associates. Besides all these educational efforts in academic, Prof. Fan has also had a remarkable impact on industry. As either previous students or scholars from Prof. Fan's group, we will talk about Prof. Fan's impacts on different areas in industry including chemicals, oil/gas, clean energy, etc. The presenters are but not limit to the following:

- Ye-Mon Chen, Shell and Quad County Corn Processors
- Ping Cai, W.R. Grace
- Bing Du, ExxonMobil Technology & Engineering Company
- Zhe Cui, Shell Global Solutions
- Luis G. Velazquez Vargas, The Babcock and Wilcox Company

217d Impact of Professor. L.-S. Fan's Scholarship Endeavors Research on Indian Chemical Society

Rushikesh Joshi

Prof. L.-S. Fan is a renowned researcher in particle science and technology, and multiphase reaction engineering with impact on many aspects of chemical engineering including energy and environmental systems as applied to such operations as resid hydrotreating and flue gas cleaning using gas-liquid-solid fluidized bed reactors, and hydrogen production from acid gas in natural gas using fixed bed reactors. He is also a highly respected scholar and educator who, besides being an excellent classroom teacher, has devoted much of his energy to mentoring undergraduate and graduate students and postdoctoral researchers and younger colleagues. The impact of his scholarship has been global and has been significant and included Indian chemical engineering community. His receipt of the prestigious AIChE Warren K Lewis Award for Chemical Engineering Education is a befitting recognition of his many outstanding contributions to education and rightfully cements his position as a top chemical engineering educator in the US.

Professor Fan's exceptional mentorship for Indian students at the PhD level has been just as remarkable as his contributions to the research field in chemical engineering. He has actively engaged with Indian students, welcoming them into his research group and advising over 40 Indian students for their PhD dissertation research. This has positively impacted the Indian community by fostering the development of future researchers in the field. This presentation will encompass the remarks garnered from both current and previous Indian students who have grown under Prof. Fan's guidance, highlighting his influence on shaping the next generation of researchers. The insights of several most influential stalwarts in Indian academia who have worked with Prof. Fan over the years for the development of a better worldwide community through education will also be included.

Professor Fan's groundbreaking innovations in clean fossil energy conversion processes have significantly reduced harmful emissions of CO2, SO2, and NOx while enabling the production of electricity, syngas, hydrogen, chemicals, and liquid fuels. His redox chemical looping technology research has been particularly impactful in India's steel industry, contributing significantly to the decarbonization of steel production through the development of efficient and sustainable processes. With his commitment to achieving a net-zero carbon future, Prof. Fan has expanded his scope globally, developing clean energy conversion technologies that have the potential to benefit the global steel industry in the fight against climate change. The impact of his work as evidenced through testimonials from experts who have witnessed the mitigation of emissions and increased process efficiency as a result of his research, will also be given in this presentation.

In recognition of his exceptional contributions to the field of chemical engineering, Prof. Fan was elected as a member of the prestigious Indian National Academy of Engineering (INAE) in 2019, in addition to holding memberships of other national academies.

Overall, this presentation highlights the significant impact of Prof. Fan's scholarships through his career long education and research endeavors on the academic and industrial development of India, particularly in the areas of environmental engineering and clean energy conversion technologies.

217e Liquid-Solid and Gas-Liquid-Solid Circulating Fluidized Beds (G-LSCFB) and Their Applications

Jesse Zhu

Although gas-solid circulating fluidized beds (GSCFB) have been around for over 70 years, liquid-solid circulating fluidized beds (LSCFB) and gas-liquid-solid circulating fluidized beds (GLSCFB) have only got the attention of researchers since the mid-1990s. Compared to the well-studied GSCFB riser, (G)LSCFB has some distinctively different characteristics, such as more uniform axial and radial flow structure, non-obvious cluster structure, a wide range of operable solids holdups, low fluid (liquid and gas) velocity and excellent mass and heat transfer. Through a series of studies in this group since 1996, as well as efforts from several other groups worldwide, fundamental understandings on LSCFB and GLSCFB have been achieved. This talk will provide a general overview on the industrial applications and potential applications of LSCFB and GLSCFB.

217f Looping and Succession for Energy Environmental Education

Liang Zeng

Professor L-S Fan is a renowned scholar and educator in the field of chemical engineering and has made significant contributions to the advancement of the discipline. He is a foreign member of the Chinese Academy of Engineering, and has significantly promote the scholarly exchanges between the United States and China. Additionally, Professor Fan has mentored a number of Chinese students and postdoctoral researchers who have gone on to pursue successful careers in academia and industry. Under his guidance, I have learned the key concepts and skills required of an independent researcher, including critical thinking, careful methodology, and technology innovation. This talk will highlight the ways in which Professor Fan's mentorship, support, and collaboration have impacted the development of Chinese students and scholars.

217g Prof. Fan As a Lifelong Mentor

Jonathan Fan

In this talk, I will discuss my experience with having Prof. LS Fan as my lifelong mentor. I will talk about how his way of teaching and mentoring has influenced my career trajectory in academia and how it has led me to be thinking about innovations that can address big societal problems.

217h A Career in Education: A Personal Reflection

Liang-Shih Fan

I am grateful to the AIChE in bestowing upon me the Warren K Lewis Award for Chemical Engineering Education. I consider this award as the most significant professional honor that I have received. This is especially the case because, many years ago, when I embarked on a career in higher education, I set the goal of achieving excellence as an educator as my foremost aim. In receiving this recognition, I take immense satisfaction in feeling that I have achieved that aim.

In pursuing this goal over all these years, I have been very fortunate to have met numerous highly distinguished scholars who mentored and supported me, guiding, influencing, and shaping me into the educator that I eventually became. They have served as exceptional role models and inspired me to assume the same role for my younger colleagues and my students. In fact, their specific impact on me has extended well beyond my pedagogical goals to influence my endeavors in research, professional service, and entrepreneurship. I am privileged to name some of them here: C. C. Chen (my BS advisor), C. Y. Wen (my PhD advisor), L.T. Fan (my post-doc advisor and my cousin), Maurice Bergougnou, Bob Brodkey, John Davidson, Norman Epstein, Daizo Kunii, Norman Li, Charlie Shao-Lee Soo, Barry Tarmy, Chi Tien, Danny I.C. Wang, Jimmy Wei, Mooson Kwauk, and Jack Zakin.

During my career at Ohio State, I have been most fortunate to have worked with a large number of undergraduate students, graduate students and post-doctoral research associates. I have found it gratifying to observe the development of their own individual careers in the years that followed their departure from my lab, whether in industry or academia. I have been privileged to mentor students at other universities where I serve as honorary professor, namely Tsinghua University in China and National Taiwan University in Taiwan. This enrichment that comes with student interaction has also excitingly extended to universities where I spent sabbatical leave including, Cambridge University in UK, Hannover University in Germany, and ETH in Switzerland. One of my goals, in addition to fostering and supporting the learning of the students in my classes, has been to combine the cultivation of the intellects and technical skills of each of my students with the assuredness of a deep commitment to an ethic of professionalism, so that each one of them will be a powerful force to the profession they serve and the society they live in. I am proud to note that many of my advisees have been very successful in academia and industry and their continued success and achievement fills me with gratitude.

I can devote much of the energy needed to pursue my career because of the full support of my wife. Our collective endeavors in exercising our shared passion for education have also affected our children in their choice of professional careers. Both my son and my daughter have established themselves in academia as professors in their respective fields.

A successful educator complements and combines inspiring teaching that effectively delivers classroom material with innovative research, scholarship and leadership, creating knowledge with insights, breakthroughs, and inventions that are ultimately embedded as new learning materials for generations to come and which benefit society at large. In receiving this honor, I am reminded once again of how deeply indebted I am to those who helped to illuminate my path as an educator desirous of impacting the discipline and the profession in a meaningful way, and to my students and younger colleagues who continually make me a better educator. I recognize all of them here with deep and abiding gratitude.

AUTHOR INDEX

AIChE
120 Wall Street, FL 23
New York, NY 10005-4020

ISBN 978-1-7138-9308-0

9 781713 893080

CURRAN ASSOCIATES INC.
proceedings.com